S0-BLM-227

THE WONDERFUL APPARITION

The Story of Halley's Comet

By Richard B. Peterson

Published by

Lighthouse Writer's Guild
P.O. Box 51277
Pacific Grove, CA 93950

First published in 1985 by Lighthouse Writer's Guild,
P.O. Box 51277, Pacific Grove, California 93950

Library of Congress Cataloging-in-Publication Data

Peterson, Richard B., 1942-
The wonderful apparition.

Bibliography: p.
Includes index.
1. Halley's comet. I. Title.
QB723.H2P48 1985 523.6'4 85-81193
ISBN 0-935-12500-0

Manufactured by Apollo Books, 107 Lafayette St., Winona, MN 55987

PRINTED IN THE UNITED STATES OF AMERICA

DEDICATION

For Adrienne,

Becky, Eric, and Michael

with much love

ACKNOWLEDGMENT

The author wishes to acknowledge the generous assistance of Ruth S. Freitag of the Library of Congress in connection with this book.

CONTENTS

Introduction

It's returning! Like a celestial sentinel, the most famous comet in history is completing its regular rounds and will soon become an object of fascination for the greater part of the human race. First detected by a specially equipped telescope in October 1982, Halley's comet was noted to be on course to present to us a heavenly display of the splendor and the wonder of the universe.

We will welcome this comet as a faithful friend that for thousands of years has returned periodically to the vicinity of the earth. Like a cosmic clock ticking off human lifetimes, Halley's comet completes each revolution in just over the biblical "threescore years and ten" allotted to each of us. This apparition will be the singular astronomical event of our lives, for only once to each generation does this celebrated comet grant an appearance.

Perhaps some moonless night far from the artificial glow of the city, we will stand with our children and gaze upon this wonderful apparition, awe-struck by its beauty and mystery. And Halley's comet will forge yet another link in a chain of generations which stretches unbroken back to a time when our ancestors fearfully viewed this very same comet from the security of a cave.

There has always been a special allure to Halley's comet. Its uniqueness as a bright, short-period comet visible to the naked eye and possessing all the cometary characteristics of its long-period brethren has made it an object of intense interest to both layman and astronomer alike. Composed of the primordial matter from which our solar system was formed, Halley's comet offers a bonanza of scientific information to an armada of spacecraft which will journey to its proximity during this return. We will learn much about the nature of these most mysterious of all heavenly bodies from the study of this particular comet.

Of perhaps even greater interest than its physical appearance and composition is this comet's long and remarkable association with people and significant historical events in Western civilization. The circumstances by which it became connected with the name of Edmond Halley justify it being one of the very few comets not named after its discoverer. Attila the Hun saw this comet prior to his defeat at the hands of the Romans at Châlons. William the Conqueror welcomed it as a sign of impending victory over Harold of England in 1066. The Angelus, the noon ringing of church bells, dates from the year 1456, when the comet hung in the sky over the head of a frightened pope. Halley's comet has been pictured as representing the star of Bethlehem and its portrait has been embroidered upon the famous Bayeux Tapestry.

Over the centuries, numerous watchers of the skies have trembled at its approach, fearing this fiery visitor either as an omen of impending disaster or as a physical threat to the earth itself. Anxiety among the populace abounded during the comet's most recent return in 1910, and similar consternation will no doubt occur during this present appearance. Halley's comet has figured prominently in mankind's perception of the heavens, and the progress of astronomy has paralleled our increasing understanding of this gypsy of the solar system, as the world's most notable astronomers, scientists, and mathematicians have sought to unlock its many secrets.

What follows is a history of Halley's comet, a celestial biography if you will. But more, it is a recounting of those individuals who have contributed to that history, and who, in so doing, have made this comet the fascinating object that it is.

CHAPTER 1

Of Conics and Comets

[An ellipse] shews a Way to the solving of the unknown walkes of comets.

—William Lower, 1610

Mankind's existence on this planet has always been punctuated by the occasional visitations of comets. Appearing suddenly in the night sky and moving swiftly across the canopy of the heavens, comets were the anomalies, the freaks, among the myriad lights of the seemingly constant stars and slowly moving planets.

The word *comet* is derived from the Greek *kometes,* meaning "long-haired," and refers to the resemblance of the long diaphanous tail to female tresses. Primitive man, if he wondered about them at all, probably held these apparitions in superstitious awe. We have learned from translated historical documents that the ancient Chinese astronomers were more objective in their celestial observations. They regarded comets as merely physical objects to be noted and chronicled but not speculated about. In the Occident, however, the appearance of a comet provoked interpretations of the event somewhere between these extreme views.

Diodorus Siculus, a Greek historian of the late first century B.C., wrote that both the Egyptians and the Chaldeans had derived methods for predicting the appearance of comets. There is no proof that they actually did so, but it is known that among the Chaldeans comets were considered to be similar in nature to the planets, but revolving in more extensive orbits about the sun, and therefore invisible most of the time. A view much like this was held by Greek philosophers of the Pythagorean school, who believed that both comets and planets were wanderers among the stars.

Among the Romans various interpretations were applied to comets. To some they represented the souls of illustrious men being carried in triumph to heaven in the form of brilliant lights. In particular the comet of 43 B.C. was viewed as a celestial chariot bearing the soul of Julius Caesar, who had been assassinated shortly before the comet made its appearance.

Both the epic poems, the *Iliad* and the *Aeneid,* contain references to comets. George F. Chambers in *The Story of the Comets* recorded translations of the relevant passages. Homer wrote that the helmet of Achilles shone "Like the red star, that from his flaming hair / Shakes down diseases, pestilence, and war." Virgil compares a hero in shining armor to a comet: "The golden boss of his buckler darts copious fires; just as when in a clear night the sanguine comets baleful glare."

Pliny the Elder (23 or 24-79 A.D.) in the second book of his *Historia Naturalis (Natural History)* classified the various types of comets according to their physical appearance, giving them such names as *barbatus,* bearded, and *monstriferus,* horror-producing (Fig.1-1). Lucius Annaeus Seneca (ca. 4 B.C. -A.D. 65), the brilliant Roman philosopher and statesman, in his *Quaestiones Naturales (Natural Questions)* came closest to a sensible view of the comets. He believed that they differed only in appearance from the planets and that they moved through similar paths in the sky. Seneca was aware of the ignorance around him, however, and believed

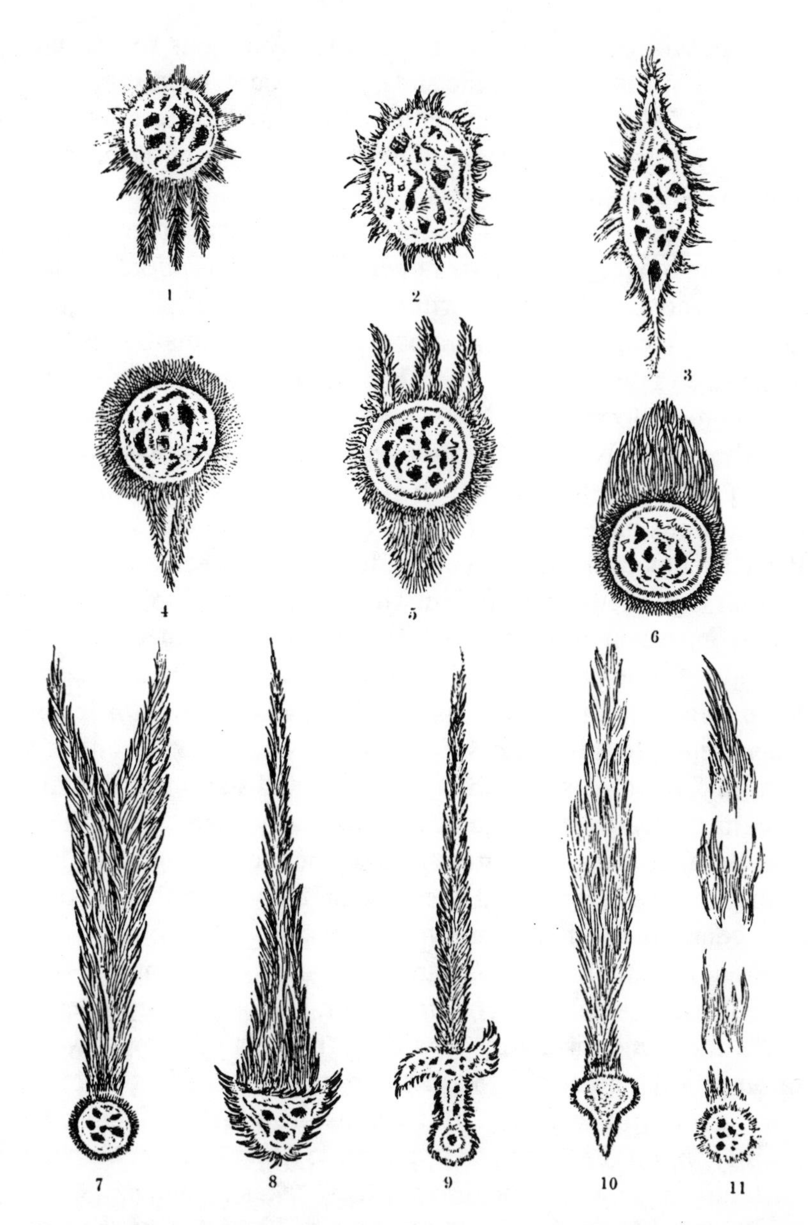

Fig. 1-1. Various Forms of Comets According to Pliny The Elder. These drawings are taken from the *Cometographia* of Johannes Hevelius which was published in 1668. When Halley's comet appeared over Jerusalem in the year 66 like a "flaming sword," it may have resembled figure 9.

"An age will come when that which is mysterious for us will have been made clear by time and by the accumulated studies of centuries." But for sixteen centuries no advance in cometary science occurred. The erroneous Aristotelian view of the cosmos effectively blocked all progress.

Aristotle (384-322 B.C.) did not believe that comets were true astronomical objects, but rather "blazing exhalations" in the upper atmosphere of the earth. All phenomena of change—rainbows, meteors, the aurora borealis, and comets—Aristotle considered inhabitants of the sublunar world and therefore not subject to uniform laws.

The first truly scientific discovery about comets involved the return of Halley's comet in 1531 and was a result of observations made by Peter Apian (1495-1552). Also known as Peter Bienewitz, Apian was a professor of mathematics at the University of Ingolstadt in Germany and a pioneer in astronomical instrumentation. His observations of the comet extended for nearly two weeks during the month of August, and he published his results in 1540 in his *Astronomicum Caesareum.* In this work Apian describes his observations of a total of five comets, including Halley's, and states that the tails of comets always appear to be directed away from the sun. This was the first substantial item of evidence toward proving that comets were not terrestrial phenomena, but were rather in some way connected with the observable universe.

A contemporary of Apian, Girolamo Fracastoro (1478-1553), an Italian physician, astronomer, and poet, independently made the same discovery regarding the tails of comets while observing the comet in 1531 and announced this fact in 1538, two years prior to Apian. But Fracastoro's lasting fame rests mainly upon his narrative poem, *Syphilis sive morbus Gallicus,* from which the dread disease takes its name.

Later, Landgrave William IV (1532-1592) of Kassel, Germany, read Apian's book. An astronomer himself, William built the first observatory with a revolving dome in 1561. He discovered the comet of 1556 and observed the comet of 1577. In both instances he found that Apian's observation applied to these apparitions. But the turning point in the history of cometary astronomy was a discovery that ultimately rebutted the prevailing Aristotelian view of the cosmos. Late in 1577, Tycho Brahe set himself to observing a bright comet which had just appeared.

Tycho Brahe (1546-1601), a giant among astronomers, was born of noble parentage at Knudstrup, Denmark (Fig. 1-2). Early in life two events galvanized his interest in observational astronomy. When he was thirteen years old, he witnessed a predicted partial eclipse of the sun. Three years later young Tycho was so impressed by the accurate prediction of a conjunction of Saturn and Jupiter that he decided to devote himself, against his family's wishes, to observing the heavens. Charles A. Whitney, in his book *The Discovery of Our Galaxy,* quoted Tycho that it was "something divine that men could know the motions of the stars so accurately that they could long before foretell their places and relative positions."

In December 1566 in a duel with another Danish nobleman, a portion of Tycho's nose was cut off. He replaced the missing flesh with a metal prosthesis. The result undoubtedly affected his social life, for he took a commoner wife. His observational skills, however, were left undiminished.

Tycho's reputation was established by celestial events that began on the evening of 11 November 1572. While taking a walk near the home of his uncle, he casually glanced up at the night sky. He saw a bright star shining in the constellation Cassiopeia, a star not evident before—a "new star" or nova. The telescope had not yet been invented, but using a sextant Tycho measured the star's angular distance from those other stars forming the constellation. He continued his observations until the end of March 1574 when the star ceased to be visible. Tycho called this phenomenon a star and not a comet, for he had determined that it lay in the region of fixed stars, whereas like

Fig. 1-2. Tycho Brahe (1546-1601). The Danish astronomer's observations of the comet that appeared in 1577 proved the extraterrestrial location of these objects.

his contemporaries he still considered comets to be generated in the upper regions of the atmosphere.

Several years later in 1576, Frederick II (1534-1588), king of Denmark and Norway, offered Tycho the two thousand acre island of Hveen in the Danish Sound as a site for construction of an observatory. Here Tycho erected Uraniborg, meaning "heavenly castle," the world's best equipped observatory at the time. It served as his home for more than twenty years (Fig 1-3).

On 13 November 1577 while fishing from one of the ponds on his island, Tycho noticed another brilliant object in the evening sky, an object with a long, broad tail that stretched across the sky—a comet. He immediately began precise measurements of the comet's changing position. Tycho was not the only astronomer to observe perhaps the most influential comet of all time, but his observations were the most complete. He saw it for the last time on 26 January 1578.

Tycho collected data from astronomers elsewhere in Europe, including Michael Mästlin (1550-1631) and William IV in Germany and Thaddeus Hagecius (1525-1600) in Prague, but Tycho was able to demonstrate more conclusively than they that for observers scattered at great distances from one another the comet appeared at any given time in approximately the same position among the stars. This lack of observable *parallax,* that is, the change of position in the sky relative to the stars when seen from different locations on the earth, indicated that the distance to the comet was considerably greater than the distance to the moon. This supralunar position of an object until then assumed to be terrestrial struck hard against the views of Aristotle. Ultimately this fact freed scientific thought from the shackles of his errant philosophy.

Not one to rush into print, Tycho waited ten years to publish his treatise on this comet, entitling it *De mundi aetherei recentioribus phaenomenis (Concerning recent phenomena of the aethereal region).* Printed on his own

Fig. 1-3. Uraniborg. Located on the island of Hveen, this was the best equipped observatory at the time in the world and Tycho Brahe's home for over twenty years. It was here that he made his precise measurements of celestial phenomena.

press, it stated the case and offered proof for the celestial position of the comet. In his later years, Brahe made extremely precise observations of the positions of the planets. It was on the basis of this information that Johannes Kepler (1571-1630) was able to derive the laws of planetary motion and lay the foundations of modern astronomy.

Brahe had established the fact that comets are true astronomical phenomena. Watchers of the skies then began to view comets with increased interest. But controversy soon arose concerning the geometry of their motion, which carried them at great speed through several constellations during their brief period of visibility. Discussion centered on their exact paths in the heavens.

Tycho Brahe ascribed to the comet of 1577 a circular path, basing this idea not on observation but on speculation. Kepler, in writing about a comet which appeared in 1607, later shown to be a return of Halley's comet, falsely conjectured that comets traversed the heavens in nearly straight lines but at varying speeds. In this same work he did guess correctly, however, that they were probably "as numerous in the heavens as are fishes in the sea."

The Welsh scientist Sir William Lower, in a letter written in 6 February 1610 to his mentor Thomas Harriot concerning Kepler's recently published *Astronomia Nova,* suggested that the "unknown walkes of comets" were greatly elongated ellipses. A similar opinion was held by Seth Ward (1617-1689), an English bishop and Savilian Professor of Astronomy. In a lecture prompted by the appearance of a comet in 1652, Ward declared that comets are "carryed round in Circles or Ellipses... so great, that the Comets are never visible to us."

The comet of 1664 provided a source for more speculation. Adrien Auzout (1622-1691), a French astronomer, attempted to predict the movements of this comet by plotting its course as a straight line lying within a single plane. His Italian contemporary, Giovanni Borelli (1608-1679), professor of mathematics at Messina and later at Pisa, who wrote under the pseudonym of Pier Maria Mutoli to avoid possible papal displeasure, suggested instead that the motion could best be accounted for not by a straight line but by a curve very similar to a parabola. This view was shared by Johannes Hevelius (1611-1687), a famous Danzig astronomer. In the first book devoted exclusively to comets, the *Cometographia* published in 1668, Hevelius suggested that comets travel in parabolic or near-parabolic orbits about the sun, but he had no proof to substantiate this idea.

It was Isaac Newton (1642-1727) who placed the matter of orbital paths of comets on a sounder basis. From his propositions regarding the motions of mutually gravitating bodies, Newton concluded in 1687 that comets, like the planets, must move under the gravitational influence of the sun, and that their paths must assume one of three geometrical curves known as conic sections.

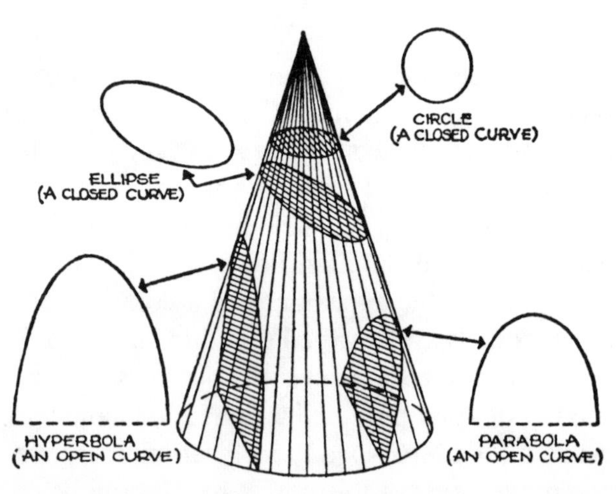

Fig. 1-4. Conic Sections. These are geometrical curves formed from the intersection of a plane with a cone. The hyperbola, parabola, and ellipse represent possible orbits of the comets.

Conics, or conic sections, are curves that are obtained when a plane intersects with a cone (Fig. 1-4). The first mathematician known to have studied them was the Greek Menaechmus (ca. 350 B.C.), a student of Plato and the tutor of Alexander the Great. The first treatise on conic sections known to have survived (seven books out of eight) was the celebrated *Conics* of Apollonius (ca. 260-200 B.C.), which was translated in 1710 from an Arabic version of the text into Latin by Edmond Halley himself. To Apollonius we owe the modern names ellipse, parabola, and hyperbola.

Conics are intersections of planes and cones, but they may also be thought of as paths of points in a plane that can be described by specific equations. Interest in conics was revived after Kepler had discovered the elliptical orbits of the planets (*Astronomia Nova,* 1609). Unlike the visible planets, however, comets are seen only during that part of their orbits when they are close to the sun. At such close range to one of the foci

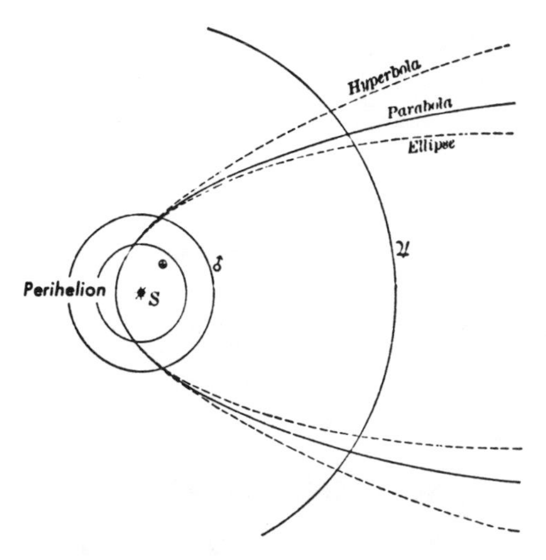

Fig. 1-5. Theoretical Paths of Comets. During the visible portion of their orbits near perihelion, comets have paths which appear to coincide.

of the various conic sections, the orbits of the comets—especially those represented by a parabola and an elongated ellipse—are very nearly congruent (Fig. 1-5). Had it been possible to track the comets out to the most distant reaches of their orbits, the problem could have been solved observationally.

Once Newton had formalized his ideas about gravity, the problem of cometary orbits was near to being solved. Newton believed, despite lack of observational evidence, that the paths of comets were parabolic, although he realized that an ellipse resulting in the eventual return of a comet was a theoretical possibility. The definitive answer could only be obtained by applying the law of universal gravitation to the specific mathematical problem of orbit computation of comets. Newton was by nature not the sort of man to attempt the time-consuming calculations necessary to do this. The laborious task fell instead to his close friend, Edmond Halley.

CHAPTER 2

Edmond Halley

From my tenderest youth I gave myself over to the consideration of Astronomy.

—Halley, *Catalogue of the Southern Stars*

Edmond Halley* is today chiefly remembered in the popular mind for the bright comet which bears his name. It was only after his cometary prediction was dramatically proven correct, however, that this scientific contribution eclipsed his other achievements, for Halley was a versatile man with many fingers in the scientific pie. During his productive lifetime Halley performed variously as an astronomer, mathematician, classical scholar, ship's captain, diplomat, editor, and publisher. Unlike many of his contemporaries, Halley kept no personal diary. His surviving letters relate almost entirely to science; of his personal life we know very little.

The correct spelling of his Christian name is "Edmond," not "Edmund," although even Halley's associates often used the latter. In the will he drew up in 1736, Halley unequivocally uses

* Not to be confused with William John Haley, Jr. (1927-1981), a singer and guitarist who as Bill Haley and his band, the Comets, pioneered rock-'n'-roll music with his 1954 recording of "Rock Around The Clock."

an *o*. As regards his surname, the pronunciation is probably "Hall´-ey," since letters to him occasionally used this phonetic form of spelling. There is little evidence to support "Hail-ey."

There is some uncertainty as to the exact date of Halley's birth. He stated that he was born on 29 October 1656 in terms of the Julian calendar used in England at the time (or 8 November according to the Gregorian calendar adopted in 1752). The records which might confirm this date, however, were presumably destroyed in the Great Fire of 1666.

The son of Edmond Halley, Sr., a prosperous landowner and soapmaker, young Halley grew up in his family's home in Haggerston, at the time a quiet hamlet, now a densely populated area in East London. He had a younger brother, Humphrey, and a younger sister, Katherine, both of whom died at an early age. According to an entry in the *Biographia Britannica* published in London in 1757, Edmond's father "resolved to give his son...and a youth of the most promising genius, an education suitable to it." Consequently, young Halley was enrolled in St. Paul's School in London, noted for its scholarship, where he became proficient in the study of both classical languages and mathematical sciences, and gained some acquaintance with astronomy.

At St. Paul's, Halley made his first recorded scientific observation, when in 1672 he measured the variation of the magnetic compass between geographical north and magnetic north. That same year his mother died. His father was later to marry a woman who would subsequently inherit a large portion of his family's wealth. At the age of seventeen Halley entered Queen's College at the University of Oxford and while still an undergraduate published his first scientific paper, the topic of which was the problem of determining the elements of the orbit of a planet. This treatise appeared in the *Philosophical Transactions of the Royal Society* in 1676. In the same year he exchanged letters with another astronomer, John Flamsteed (1646-1719), the first Astronomer Royal (Fig. 2-4).

Flamsteed wrote highly of the young astronomer to a friend, Richard Townely: "I have met also with an ingenious youth versed in calculations and almost all parts of mathematics, tho yet scarce 19 years of age; Mr. Edmond Halley, whose assistance I hope to have often..." This early friendship later dissolved into bitterness and enmity, due in large measure to their utterly different personalities. Flamsteed was a methodical astronomer as well as a clergyman. He came to resent Halley's flamboyant nature and frivolous attitude toward the Bible. According to the *Biographia Britannica* , Halley possessed "a vein of gaiety and good humour" and "always spoke as well as acted with an uncommon degree of sprightliness and vivacity." It was also noted that he was a man "of middle stature, inclining to tallness, of a thin habit of body, and a fair complection."

Very early Halley's thoughts were directed toward compiling a catalogue of stars so that precise reference points for the tracking of the planets could be obtained. He knew that such catalogues were being prepared by Hevelius at Danzig, Flamsteed at Greenwich, and J.D. Cassini at Paris, and so he planned to chart the southern stars invisible from Europe.

At the age of twenty, Halley voluntarily left Oxford without a degree to journey to the island of St. Helena, which at that time was the southernmost land under England's flag. This island became famous a century and a half later as the residence of the exiled Napoleon Bonaparte from 1816 to 1821. Financed by an annual expense account of three hundred pounds provided by his father, Halley obtained permission and free passage from the East India Company, which had jurisdiction over the island. Accompanied by a friend, Mr. Clerke, of whom we know little else, he set sail in November 1676. The two men landed in February 1677 and set up their observatory on a central prominence, known to this day as Halley's Mount.

The living conditions in the isolated interior of the island were primitive, and Halley got along poorly with the Governor, Gregory Field, who was both discourteous and arrogant. Halley

Fig. 2-1. Edmond Halley As a Young Man. This portrait was probably painted by Thomas Murray in the last decade of the seventeenth century as Halley was approaching forty years of age. The inscription was added at a later date. Halley holds his graphical solution for determining the roots of a biquadratic equation.

had selected this site partly on the basis of favorable weather reports related by travelers who had touched at the island. But he was disappointed and frustrated by cloudy weather and found the observing conditions to be even more unfavorable than in England. Halley made use of every moment of clear nighttime sky, but succeeded in determining the positions of only 341 stars of the Southern Hemisphere, six of which he obtained from his observations made at sea.

After a year on the island, Halley returned to London in May of 1678 and immediately made arrangements for the publication of the results of his observations. The work was issued late in 1678 under the title *Catalogus Stellarum Australium (A Catalogue of Southern Stars)* . Each star was arranged by constellation and was listed by magnitude and its celestial longitude and latitude. This publication secured Halley's reputation. Flamsteed acclaimed him as "Our Southern Tycho." The Royal Society elected him a Fellow in recognition of the contribution he had made to astronomy, and at the intercession of King Charles II he received an honorary Master of Arts degree from Oxford. Halley was just twenty-two years old.

Within a year after his return from St. Helena, Halley was chosen by the Royal Society to represent them in a dispute between Johannes Hevelius and Robert Hook. The international controversy concerned the best method to be used in the determination of star positions. Hevelius preferred observation with the naked eye aided by sights, while both Hook and Flamsteed advocated the use of the telescope.

Halley, who was chosen as the arbiter because of his great natural diplomacy, arrived in Danzig in May 1679 and was warmly welcomed by Hevelius. Halley stayed with the great astronomer for two months and made observations with both his own and Hevelius' instruments. Halley could find no significant difference in accuracy between the two methods, although later studies did prove that greater accuracy of position was possible

Fig. 2-2. Edmond Halley in the Uniform of a Naval Captain. This is the best known portrait of Halley and was painted by Godfrey Kneller around 1700.

by using the telescopic sight. As a result of this meeting, both Halley and Hevelius remained lifelong friends.

After his return to London, Halley lived for a time with his father. In December 1680, he embarked upon a grand tour of the Continent in the company of his close friend, Robert Nelson (1656-1714), a religious young man well known at the time for his devotional writings.

At the time of his departure, Halley was intrigued by a bright comet visible to the naked eye, which he had first noticed the previous month. The comet reappeared during the journey to Paris and for a time was thought to be a different object. Comets were still a mystery to astronomers in 1680 and the appearance of this particular comet was bound to stimulate a man of Halley's astronomical interest in such apparitions.

Late in 1681 while traveling in Italy, Halley was called home, possibly because of financial difficulties which were beginning to beset his father. He returned to London on 24 January 1682. Within three months of his arrival, he married Mary Tooke, a daughter of the auditor of the Exchequer. According to a quote in the biography of Halley by Colin A. Ronan, she was "a young lady equally amiable for the gracefulness of her person and the beauties of her mind..." They lived happily together until her death 55 years later.

There were three surviving children from their marriage, a son Edmond, who became a naval surgeon and who died about two years before his father, and two daughters, Margaret, who died a spinster at age 55, and Catherine, twice married and died at age 77.

The young couple set up house at Islington, which then lay on the northern outskirts of London. There Halley established and equipped his own private observatory and commenced a series of observations of the moon and planets. He intended to correct existing tables of the moon's motion so that these tables might ultimately be used to determine longitude at sea, a most pressing navigational problem of his day.

His work went smoothly until it was interrupted by a family crisis. On the morning of Wednesday, 5 March 1684, Edmond Halley, Sr. left home, leaving word that he would return that evening. He was not seen alive again. Five days later his body was recovered from a nearby river. Although foul play was suspected, no assailant was ever found.

At about this time Halley had been pondering Kepler's laws of planetary motion. Halley had determined that the sun's attraction upon the planets decreased as the inverse of the square of the distances between them, but he could not deduce an orbit which would match the observed motions of the planets. So in August 1684 Halley journeyed to Trinity College, Cambridge, in order to pursue the matter further with Isaac Newton, who at age forty-one already had a reputation as a brilliant mathematician (Fig. 2-5). Halley asked Newton, "If the inverse square law be true, what will be the path of a planet?" Newton replied without hesitation, "An ellipse." When Halley asked how he could be so sure of this, Newton replied, "Why, I have calculated it." Three months later Newton sent Halley the written proof he had wanted.

Impressed by the significant ideas and discoveries of Newton, Halley encouraged the naturally shy man to expand his studies on celestial mechanics, with the aim of eventual publication. The ultimate fruition of the association between these two men was the *Philosophiae Naturalis Principia Mathematica (Mathematical Principles of Natural Philosophy)*. The *Principia* was completed by Newton in a brief fifteen months—sometimes described as the greatest concentration of mental effort ever made by one man.

Halley seems to have been having personal financial difficulties at this period, probably as a result over litigation of his late father's estate with his stepmother. He accepted a salaried position on 27 January 1686 as Clerk to the Royal Society and editor of its *Philosophical Transactions*. To

Fig. 2-3. Edmond Halley At Age 80. This portrait is by Michael Dahl and was painted in 1736. Halley holds a diagram of his geomagnetic hypothesis.

this publication he contributed many original investigations and discoveries over the next thirty years.

While Newton was busy preparing his monumental treatise, Halley was arranging for its printing and publication. The Royal Society had initially agreed to publish the work, but had in the meantime become enmeshed in the publication of a work by Francis Willoughby entitled *Historia Piscium (History of Fishes)* . This volume had not sold well and had involved the Society in considerable loss. The Society was left with a large number of copies of the book and even found it necessary to pay Halley for his duties as Clerk by giving him 75 copies of the *Historia* in lieu of a portion of his salary.

Eventually Halley provided the financial support for the *Principia* out of his own pocket and saw it published in July 1687. Augustus De Morgan (1806-1871), a historian and biographer of both Newton and Halley, summarized Halley's achievement when he wrote, "But for him, in all human probability, the work would not have been thought of, nor when thought of written, nor when written printed."

In the preface to the first edition of the *Principia* , Newton refers to "the most acute and universally learned Mr. Edmond Halley." Indeed, Halley's scientific versatility during this period was astounding. In 1686, he published a map of the world which showed the distribution of prevailing winds over the oceans, the first meteorological chart to be published. In 1693, he constructed mortality tables for the city of Breslau, one of the earliest attempts to relate mortality and age in a population. This study laid the foundations for the future development of the actuarial tables used in life insurance. He wrote about such diverse topics as ancient measures of weight, a remedy for a disease of the skin, the growth of trees, identification of Roman towns in Great Britain, and the microscopic examination of crystals.

In 1691 when Halley was thirty-five, a position as Savilian Professor of Astronomy at Oxford fell vacant, a professorship

Fig. 2-4. John Flamsteed (1646-1719). The first Astronomer Royal, he was succeeded by Halley in this post. Flamsteed resented Halley's flamboyant nature and unorthodox religious views and continually sought to discredit him.

established through an endowment by Sir Henry Savile (1549-1622), an English mathematician and tutor to Queen Elizabeth. Halley was an eager candidate for the position, for it would provide him with leisure time that would enable him to pursue his research activities, including the study of comets. The appointment, like all Oxford professorships at the time, required Church approval and general assent of the candidate to the Book of Common Prayer. Religiously Halley was a freethinker, who rejected the literal interpretation of the Bible. Specifically, Halley's own speculations regarding the age of the earth differed from the 4,000 years assumed by seventeenth-century biblical scholars and by Newton, an avid lay scholar. The taciturn Flamsteed had opposed Halley's election on the grounds that he would "corrupt ye youth of ye University with his lewd discourse." Halley's heterodox religious opinions, together with Flamsteed's continued animosity, were the likely reason the appointment went to a Scotsman, David Gregory (1661-1708).

In 1696, at the recommendation of Newton, Halley was appointed Deputy Comptroller of the Mint at Chester. Defacement of the silver coinage by clipping small pieces from them had become a major problem, and it was Halley's job to supervise the manufacture of silver coins with milled edges to safeguard them from such vandalism. Although unhappy at this job, Halley remained at it until closure of the mint two years later. During this time Halley retained his position at the Royal Society, reporting on various matters of scientific interest.

At about this time the twenty-six-year-old Tsar of Muscovy, known to history as Peter the Great (1672-1725), was in England to study British shipbuilding. Because of Peter's interest in scientific matters, Halley was his frequent guest. The two men enjoyed stimulating conversation and companionship. Peter was a man accustomed to enjoying life to the fullest, and anecdotes abound concerning the Tsar's wild behavior. One such alleged incident occurred in the company of Halley. After a day of discussion, an early dinner, and much wine, the two men went for a walk on the elegant grounds of Sayes Court, where Peter was temporarily staying, the country home of John Evelyn (1620-1706), the diarist and horticulturist. Halley is said to have given Peter a wheelbarrow ride that ended in one of Evelyn's prized holly hedges. This account may be apocryphal, but the damage caused to Sayes Court by the Tsar and his retinue, amounting to over three hundred pounds, is not.

On 4 June 1696, Halley was commissioned as a naval captain of one of His Majesty's ships, perhaps as a result of his experience with ocean navigation during his voyage to St. Helena. He was to command this ship on the first sea voyage undertaken for purely scientific purposes. Setting sail in the *Paramore* , a three-masted, flat-bottomed ship known as a pink, his orders instructed him "...to Improve the knowledge of the Longitude and Variations of the Compasse..." Halley hoped that by charting *magnetic variation,* the departure of the compass needle from true north, he could determine longitude at sea. His voyages traversed the vast Atlantic Ocean, as far south as the Falkland Islands, and lasted for two years. Despite the spartan life aboard ship and a series of misadventures that included having his ship mistaken for a pirate vessel and fired upon by two English merchantmen, suppressing the mutiny of his first lieutenant, navigating the ice fields of the South Atlantic, and suffering a pestilential disease, Halley successfully commanded his vessel and returned with the scientific information he needed for the completion of his chart of the variation of the compass.

In 1701, Halley published the scientific result of this voyage, the Atlantic Chart, the first to adopt isogonic lines (or "Halleyan lines" as his contemporaries called them) to connect points of equal magnetic variation. Such lines have been used ever since for charts of this kind.

During part of 1702 and the year following, Halley was sent at the behest of Queen Anne on two diplomatic missions to Vienna. The object was to organize seaports on the Adriatic and

Fig. 2-5. Isaac Newton (1642-1727). This portrait represents him at age 46. The friendship and scientific cooperation between Halley and Newton proved beneficial to them both.

to advise the Emperor Leopold on the fortifications of Trieste, a subject at which Halley had previously shown himself adept when, during his survey of the English channel in 1701, he provided intelligence reports on French port fortifications.

In the autumn of 1703, John Wallis, the Savilian Professor of Geometry at Oxford, died, and Halley again was an applicant. Flamsteed noted that "...Mr. Halley expects his place, who now talks, swears, and drinks brandy like a sea-captain; so that I much fear his own ill-behavior will deprive him of the vacancy." However, early in 1704 Halley was elected without dissent and honored with the degree of Doctor of Laws. At last Halley was able to look forward to a time of comparative leisure, and at once he directed his efforts to comets, an interest that had lain dormant for several years.

In 1705, he published his *Astronomiae Cometicae Synopsis (Synopsis of Cometary Astronomy)*, in which he predicted the return of the comet of 1682, later to be named Halley's comet. The first study to establish a comet's periodicity, this is the scientific discovery to which Halley owes his fame.

He continued with his diverse projects, including a translation from Arabic into Latin of the *Conics* of Apollonius of Perga, one of the greatest mathematicians of antiquity. In 1716, he devised a method for observing transits of the planet Venus across the disk of the sun, in order to accurately determine the distance of the earth from the sun. In 1718, Halley was the first to show that the stars were not fixed in the heavens, but that at least three bright ones—Arcturus, Procyon, and Sirius—had actually changed their positions since the time of the Greeks.

In 1713, Halley was appointed Secretary to the Royal Society, and on 9 February 1720, following the death of the Astronomer Royal John Flamsteed, Halley was named his successor. Halley began a long series of observations directed mainly to timing the transits of the moon across the meridian and providing new data to improve existing lunar tables useful in determining the longitude at sea, so necessary for perfecting the art of navigation. Although by 1731 he was able to determine longitude with an error of no more than sixty-nine miles at the equator, the eventual solution to the problem awaited the invention of an accurate marine chronometer by John Harrison (1693-1776), a young, self-trained clockmaker.

Because Halley was sixty-four when he was appointed Astronomer Royal, his subsequent scientific production was unremarkable. As De Morgan later commented, "The period during which he held the post of Astronomer Royal, compared with those of his predecessor Flamsteed and his successor Bradley, is hardly entitled, if we look at its effect upon the progress of science, to be called more than a strong twilight night between two bright summer days."

Prior to the weekly meetings of the Royal Society, Halley would journey up to a London coffeehouse and dine with his friends. These informal gatherings had been instigated by Halley and gradually evolved into the Royal Society Club, formally constituted in 1743, the year after Halley's death. According to Sir Joseph Ayloffe, who was a member of this group, it was the custom to order fish and pudding with their beer since "Dr. Halley never eat any Thing but Fish, for he had no Teeth."

On 30 January 1736, Mary Halley, with whom Halley had lived contentedly for over fifty-five years, died. Shortly thereafter, Halley suffered a paralysis in his right hand, the result of a stroke. With the aid of an assistant, however, he was able to continue his observations. In 1741, Halley's son died and soon after his own health began to fail. According to the *Biographia Britannica* , "...his paralytic disorder gradually increasing, and thereby his strength wearing, though gently, yet continually, away, he came at length to be wholly supported by such cordials as were ordered by his Physician." It was on 14 January 1742 (or 25 January by the Gregorian calendar) that "...being tired...he asked for a glass of wine, and having drunk it presently expired as he sat in his chair without a groan, ...in the 86th year of his age."

At his own request Halley was laid to rest with his wife in St. Margaret's churchyard at Lee, not far from Greenwich. The epitaph upon his tomb, translated from the Latin, reads in part, "Under this marble peacefully rests, with his beloved wife, Edmundus Halleius, LL.D., unquestionably the greatest astronomer of his age."

CHAPTER 3

The Prediction

I may venture to foretell that it will return again in the year 1758.

—Edmond Halley

In the year 1680, a great comet appeared in the skies over Europe. First discovered by the German astronomer Gottfried Kirch (1639-1710) at Coburg, in Saxony, on 14 November in the constellation of Leo, this comet caused great alarm. Fear of impending disaster and approaching doom was common among the populace. The Church had medallions made to be distributed by monks as amulets to ward off the malign influence of the comet (Fig. 3-1). A motto was written upon each medallion. The translation from the German reads, "The star threatens evil; trust in God who will make things turn to good." If the large Roman capitals are read as numerals and added together, they make 1681. During the first two weeks of January in 1681, the comet was at its brightest.

When this comet appeared, Edmond Halley was twenty-four years old and his friend Isaac Newton had just turned thirty-eight, still seven years away from publishing his monumental *Principia*. It would not be until a quarter century later that the events surrounding the comet's appearance would culminate in Halley's famous prediction.

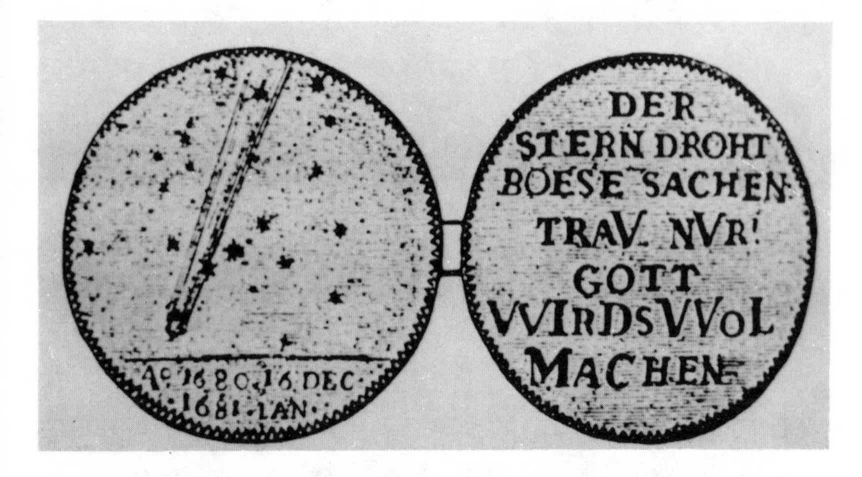

Fig. 3-1. The Comet of 1680 Pictured on a German Medallion. The appearance of this comet aroused Newton's interest in these objects and played an important part in the studies which later culminated in his theory of comets in 1687.

Georg Samuel Dörffel (1643-1688), a pastor in Plauen, Germany, was the first to describe the path of this comet as a parabola with the sun at the focus of this curve. His hypothesis, however, received little notice at the time. The *perihelion* passage, or closest approach to the sun, took place on 18 December. The comet was now lost in the sun's glare. Upon reappearance, the comet displayed a tail that extended in an arc 70 degrees to 90 degrees in length and curved like a sabre. During the early part of January 1681, this brilliant comet was observed throughout the world. Using his seven-foot telescope, Isaac Newton was the last person to witness this apparition on 19 March.

Newton's observations of this comet revived a previous interest in astronomy. His subsequent conjectures based on these observations embroiled him in an extended debate with John Flamsteed, the Astronomer Royal. In December 1680, the comet passed out of view from the earth. After perihelion it became visible again. The question was whether this was the same comet or another one. Newton thought that the pre-and post-perihelion apparitions were two different comets. In a letter to Flamsteed dated 28 February 1681, he wrote, "I am further suspicious that the Comets of November and December...were two different ones." Flamsteed, however, believed that the two apparitions were in fact the same comet.

It was several years before Newton came around to Flamsteed's interpretation of this event. In a letter of 19 September 1685, he wrote to Flamsteed, "I have not yet computed the orbit of a comet, but am going about it, and taking that of 1680 into consideration, it seems very probable that those of November and December were the same comet." This understanding of the behavior of a comet subsequently aided Newton in developing the law of universal gravitation.

Meanwhile, another comet had appeared in 1682. Although it did not appear as bright as its predecessor of two years previous, it too caused great alarm in many parts of Europe and handbills urging people to repentance were distributed (Fig. 3-2). This was the comet whose periodic returns were to be predicted by Edmond Halley and which subsequently would bear his name.

The first recorded observation of this appearance was by the pastor of Plauen, Georg Dörffel, on 15 August. The comet was also detected at the Royal Observatory in Greenwich by Flamsteed's assistant while he was scanning the northern heavens with a telescope. By 21 August the tail was noted to be 10 degrees long. Jean Picard (1620-1682), the father of modern astronomy in France, sighted this comet on 26 August and continued to observe it until 12 September, one month before his death. Johannes Hevelius of Danzig said that by the end of August the comet was bright and could be seen throughout the night with a tail 12 degrees to 16 degrees long. About 8 September a kind of luminous ray burst from the nucleus into the tail. Hevelius was impressed by this phenomenon and drew

Eigendliche Vorstellung und wahrhafte Beschreibung

Des in Hungaren und andern angränzenden Ohrten beobachteten erschrekenlichen Luftgesichtes und erstaunlichen

Wunder-Sternens.

Großgünstiger / Geehrter Leser!

Es wird dir allhier unter Augen geleget ein erschrekliches Luftgesicht / mit welchem der Allgewaltige Gott abermahl neben noch andern Gesichtern (wie wahrhafter Bericht faller) das liebe Hungerland erschreket und zu gleich die ganze Christen-Welt zur Busse aufwecken wollen. Das erste mahl/das es diser Welt erstaunlich erschienen ware/geschahe den 1. Tag Jenner des mit Gott angefangenen 1682. Jahrs/über der Vestung Neuhäusel; Hernach den 10. Tag und etliche nachfolgende Tage im erlittenem Hornung dises Jahrs/Morgens früh um 4. Uhren zu Leopoldstatt über dem Galtgozienser Gebirge auf/den Sternen gegen Mähren/den Schweiff aber gegen der Türkey wendende. Der Sterne war sehr groß und hell/nicht feuerig/sonder weißliche/gleich dem Liechte des Mondes; Der Schweiff war krum und gleichsam Schlangenartig gewunden wie der Bliz in sich zeigend: einige hin und weder ob sich und nid sich steckende Pfeil-Spitzlein und zu Ende etwas / das sich einem Türkischen Federwadel gleichete; Der Schweiff endete sich in siben Pfeilspitze/die sich gegen der Türkey außbreiteten. Oben auf dem Ende des Schweiffes liesse sich eine Kron sehen/gleich wie in der Mitte darunter eine andere/die auß den Wolken heraußschimmerte. Nebendhin erschienen 2. Türken Köpfe und einige Mondsgesichter/die theils Kugelförmig waren. Die Länge des ganzen Zeichens beluffe sich/dem ungefährlichen Augenmosse nach zu rechnen/etwann auf 20. Ellen; Dessen Währung aber war durch de Glanz der herfür blinkenden Sonnen nach weniger Zeit unterbrochen / und dardurch das ganze Gesichte den Augen entzogen. Ist gewüßlich ein erschreklicher Aufwecker zum Neuen Jahre gewesen; Dessen Deutung wir dem Allwissenden Gott heimsetzen/der uns auß seiner Genade wachsame Herzen geben/die Plag von unsern Hütten abwende und hingegen die bedrohenden Pfeile über seine und seiner Kirchen Feinde außsenden wolle!

Gott warnet nicht umsonst durch Luft und Schreckens-Zeichen;
Wer denen widerstrebt und sich nicht kehrt darn/
Geht immer fort mit Schand auf schnöder Laterbahn/
Und wird des Höchsten Gnad niemahlen mehr erreichen/
Weil Er auf Gnade hin hat immer mißgethn/
Merk drauf/o lose Welt! und stosse dich dara:
Laß dise Wunderding Dich in der Buß erweichen!

Fig. 3-2. Handbill Describing the Appearance of Halley's Comet in 1682 and Urging Repentance. The Turkish symbols reflect the great alarm caused by the Turkish invasion of Europe. The only known copy of this handbill is contained in the library of Pulsnitz Observatory in East Germany.

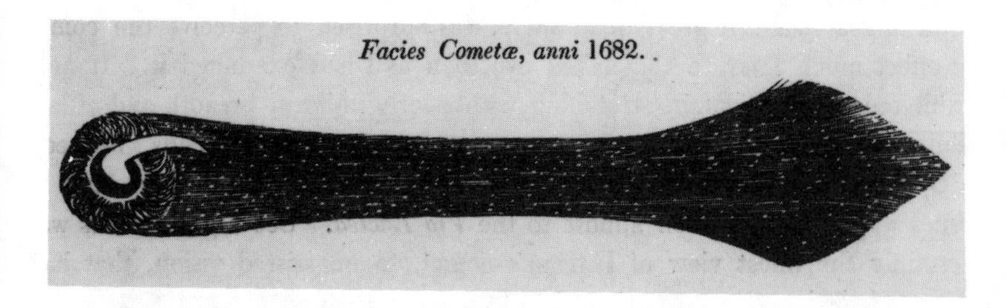

Fig. 3-3. Halley's Comet in 1682 As Drawn by Johannes Hevelius in His *Annus Climactericus*. The luminous ray rising from the nucleus is shown. A similar event occurred during the comet's return in 1835.

a picture of it (Fig. 3-3). A similar event occurred during this comet's return in 1835. Such a flare indicated violence within the comet's internal structure. Astronomers also noted that the tail was not directed exactly away from the sun as with previous comets and that the comet was moving in a retrograde path—that is, opposite to the normal direction of motion of the planets—and slightly inclined to the *ecliptic*, the plane of the orbit of the earth around the sun. As the comet neared the sun and passed perihelion on 15 September, it approached the Southern Hemisphere and was lost to view.

Edmond Halley also watched this comet. His original observations were not discovered, however, until early in this century. In a college notebook upon which he had written "Edmund Halley his Booke and he douth often in it Looke," his observations and calculations were carefully noted in Latin (Fig. 3-4). These naked eye observations, extending from 26 August through 9 September, were crude determinations based upon alignments with stars. Several years later when Halley set about to determine the orbit of this comet, he did not base his computations on his own inaccurate sightings but rather upon the very accurate observations made by Flamsteed.

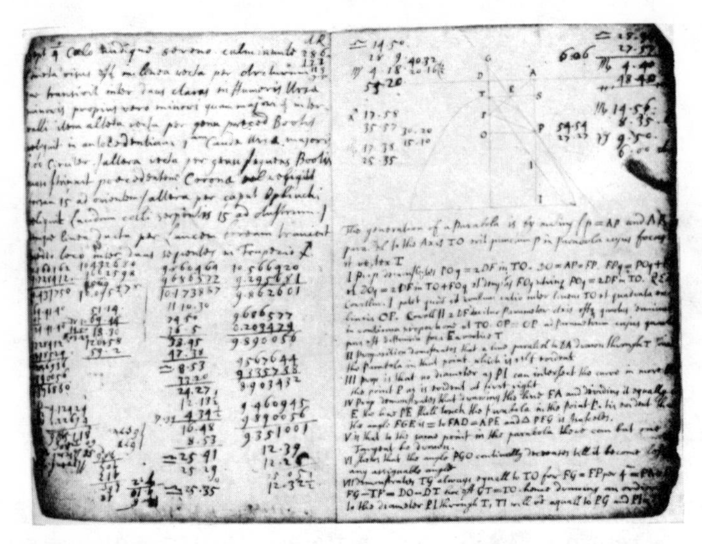

Fig. 3-4. Halley's Observations of His Comet on 4 September 1682. The left-hand page from his notebook contains his observations with their reductions; the right-hand page by a curious coincidence contains notes on the parabola.

On 6 April 1687, the third book of Newton's *Principia, De Systemate Mundi (The System of the World)* was presented to the Royal Society of London. It was noted in a later report of this meeting that the work "contained the whole system of celestial motions, as well as the secondary and primary planets, with the theory of comets." It was in this work that Newton stated the law of universal gravitation. His description of the orbits of comets was that they moved in conic sections and that those comets that return move in elliptical orbits and follow Kepler's third law but that "their orbits will be so near to parabolas, that parabolas may be used for them without sensible error." "But," he said, "I leave to be determined by others the transverse diameters and periods, by comparing comets which return after long intervals of time in the same orbits."

Newton had explained the method of determining by geometrical construction the visible portion of the path of a comet. It was Edmond Halley who undertook the laborious task of calculating orbits to determine whether any previously recorded comets appeared to follow an identical path and might therefore be considered periodic. Because of his many and varied interests, Halley did not begin the mathematical work until 1695.

Using Flamsteed's observations, Halley began to calculate the parabolic elements of the comet of 1682 in accordance with the methods devised by Newton. In a letter to Flamsteed on 14 September 1695, Newton stated, "Mr. Halley was with me about a design for determining the orbs of some comets for me. He has since determined the orb of the comet of [1682] by my theory; and finds, by an exact calculus, that it answers all your observations and his own to a minute."

Acting on the advice of Newton, Halley made a diligent search among past historical records of comets for those where sufficient information might be found for computing their orbits. All those comets appearing prior to the fourteenth century had not been observed adequately enough to ascertain their orbits. But beginning with the comet of 1337, Halley eventually assembled a total of twenty-four comets of which sufficiently precise observations had been made and recorded to allow him to calculate the actual paths they followed when visible. In addition to the comet which he observed in 1682, Halley found those which appeared in 1531 and 1607 to be of particular interest.

The comet of 1531 was first observed at the end of July and remained visible until early September. It was seen throughout Europe and was noted to be of a reddish or yellow color. Peter Apian closely observed this comet and he described its appearance and characteristics in his *Astronomicum Caesareum.* His sightings extended from 13 to 23 August by the Julian calendar and were sufficiently accurate for Halley's purposes.

During the latter part of September in the year 1607, a comet was observed one evening by Johannes Kepler at Prague while he was returning from a party. The same comet had been detected and recorded by a monk in Swabia several days before.

To Kepler the comet had the appearance of a star of the first magnitude and, as best he could see, was without a tail. Kepler followed the comet during September and October and published his observations in Augsburg in 1619 in a work entitled *De Cometis libelli tres*. Both Christian Severin (1562-1647), known as Longomontanus, a disciple of Tycho Brahe, at Copenhagen and Thomas Harriot (ca. 1560-1621), an English mathematician and astronomer, also observed the comet. According to Longomontanus, the tail was of considerable length and more dense than usually seen in comets. It was also noted to fluctuate in appearance. The head of the comet, to the naked eye, appeared to be the size of Jupiter and its color similar to that of Saturn. The comet passed through Ursa Major, Boötes, Serpens, and Ophiuchus and disappeared from view during early November. Although Harriot's observations of this apparition were later shown to be the most accurate, Halley used Kepler's sightings in his computations.

By late 1695 Halley suspected the comet of 1682 to be periodic, for as he wrote to Newton on 28 September of that year, "I am more and more confirmed that we have seen that Comett now three times, since ye Yeare 1531." But other scientific pursuits and official duties competed for his time, and it was not until 1704 when he became Savilian Professor of Geometry at Oxford that he had the relative leisure to complete his calculations of cometary orbits.

Halley had to define the orbit for each of the twenty-four comets for which he had adequate observations, since on the basis of appearance alone one comet may be indistinguishable from another. At least three observed positions of the comet were, according to Newton, necessary to determine its *elements* or complete orbital description. Each observation consists of two celestial coordinates, the declination and right ascension, which correspond to latitude and longitude on the earth.

The *celestial sphere* is an imaginary sphere, with the earth as its center, upon which the heavenly bodies appear to be projected (Fig. 3-5). The *celestial equator* is the intersection of the plane formed by the earth's equator with the celestial sphere. The two *celestial poles* are continuations of the line of the earth's axis extended into infinity. A system of equatorial coordinates has been devised similar to the geographical system used on earth. *Declination* (denoted by δ, the lowercase Greek letter delta) represents celestial latitude measured in degrees either north or south of the celestial equator. *Right ascension*, (α, the lowercase Greek letter alpha) is analogous to geographical longitude. The celestial equivalent of earth's zero meridian at Greenwich is the direction to the *vernal equinox,* which is the line extended from the center of the sun's disk to the constellation of Aries. Spring begins in the Northern Hemisphere when the sun enters this constellation as seen from earth. Right ascension is measured eastward from the vernal equinox and is expressed in hours, minutes and seconds, in which 24 hours is equal to

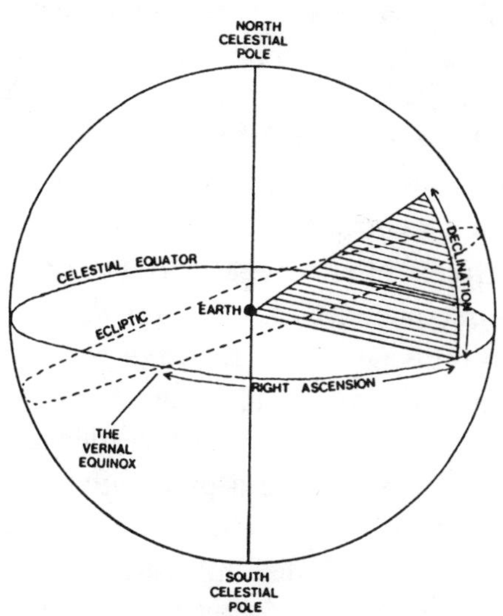

Fig. 3-5. The Celestial Sphere with the Equatorial Coordinate System. The celestial equator is an extension of the plane of the earth's equator and is inclined nearly 23-1/2 degrees to the ecliptic. Celestial latitude is referred to as declination and celestial longitude as right ascension.

360 degrees. Thus, specifying declination and right ascension defines a unique point on the celestial sphere.

Once a comet is located on the celestial sphere as seen from the earth—its apparent position—the next step is to calculate its position in reference to the sun, the actual center of its motions. In this heliocentric system of reference, positions are located by the *heliocentric latitude* and *heliocentric longitude*. Latitudes are measured not from the celestial equator but from the ecliptic. The zero longitude remains the vernal equinox. The computations that convert earthbound observations of an object's apparent position in the sky into its actual position in the heliocentric system are called the *reduction* of the observations. This was Halley's first task.

Next, Halley had to describe the complete orbit of the comet by obtaining the elements of that orbit. He did some very original work in this regard, and astronomers today still use his general scheme.

A total of six quantities completely describe the orbit of a comet. Three angles describe the orientation of the orbit in space (Fig. 3-6). The *argument of perihelion* (designated by ω, the lowercase Greek letter omega) has replaced the obsolete term called the heliocentric longitude of perihelion. The other two angles are the heliocentric *longitude of the ascending node* (Ω, the uppercase Greek letter omega) and the *inclination* (i) of the orbit to the ecliptic plane. The *perihelion distance* (q) measured in astronomical units (AU) and the *time of perihelion passage* (T) further define the size and form of the orbit. The sixth element, the *eccentricity* (e), measures the departure of an ellipse from circularity. By assuming a parabolic orbit, Halley could ignore this element and simplify his computations.

Finally, after what Halley himself termed "prodigious" labor, an orbit was calculated for each of the twenty-four comets (Table 3-1). Halley was excited to discover that several orbits were similar. This suggested one of two possibilities: different

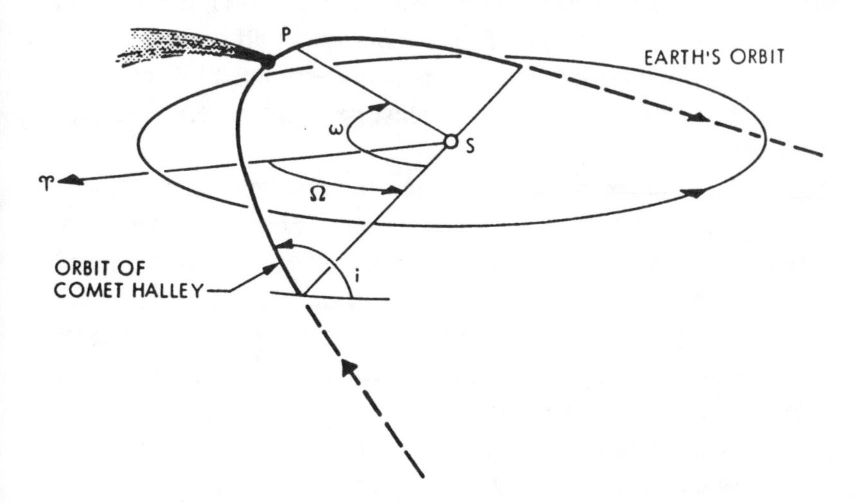

Fig. 3-6. Angular Elements of the Orbit of Halley's Comet. These three elements include the argument of perihelion (ω), the longitude of the ascending node (Ω), and the orbital inclination (i). The direction to the vernal equinox is shown. The line SP represents the comet's distance to the sun at perihelion. The broken line indicates that portion of Halley's orbit below the ecliptic plane.

comets follow the same path at nearly equal intervals of time, or a single comet has an elliptical orbit, which in the region of the sun would differ little from a parabolic one. He noted that the three comets of 1531, 1607, and 1682 all moved in retrograde orbits. Their orbital periods, however, were not precisely equal, and the inclinations of their orbits were not identical.

Between the perihelion passages of 1531 and 1607 there had been a time interval of 76 years and 6 weeks, while the time between the passages of 1607 and 1682 had been only 75 years less about 5 weeks. Halley theorized that the gravitational attractions of the major planets, especially Jupiter, might disturb the comet's orbit (called a *perturbation*) and that this might account for the discrepancies in its interval of return and orbital inclination. He noticed that the comet's close approach to Jupiter during its most recent return had probably shortened its period, and surmised that its next appearance would probably be delayed until sometime late in 1758. He could not predict the disturbing effect more accurately since mathematical methods for calculating the effects of perturbations had not yet been developed. As it turned out, Halley either estimated the influence of Jupiter fairly shrewdly, or at least he made a very fortunate guess.

In 1705 Halley finally presented the results of his cometary research in a memoir to the Royal Society. Entitled *Astronomiae Cometicae Synopsis (A Synopsis of the Astronomy of Comets)* and published in the *Philosophical Transactions* for March 1705, this report, printed in fairly large type, fills less than eighteen small quarto pages. But it established Halley as the first great calculator of comet orbits. In it he wrote:

> And, indeed, there are many things which make me believe that the comet which Apian observed in the year 1531 was the same with that which Kepler and Longomontanus more accurately observed in the year 1607, and which I myself have seen return in the year 1682. All the elements agree, and nothing seems to contradict this my opinion, except that there is an inequality in the times of revolution, but this is not so great that it cannot be attributed to physical causes. For example, the motion of Saturn is so disturbed by the other planets, and especially by Jupiter, that its periodic time is uncertain to the extent of several days. How much more liable to such perturbations is a comet which recedes to a distance nearly four times greater than Saturn, and a slight increase in whose velocity would change its orbit from an ellipse to a parabola.... Hence I think I may venture to foretell that it will return again in the year 1758.

At first Halley ventured to express his opinion concerning the periodicity of the comet merely as a conjecture, for the several orbital discrepancies still worried him. He also noted that the orbits of the comets of 1661 and 1532 were similar, and he erroneously suggested in this same treatise that these two comets were the same object, with a period of 129 years and an expected return date of about 1790. But no comet following a similar path appeared at or near that date.

Halley also believed, along with Newton, that the bright comet of 1680 had an orbital period of 575 years, with previous returns in 1106, 531, and 43 B.C. This misconception was also shared by Newton's clergyman friend, William Whiston (1667-1752).

Whiston succeeded Newton as Lucasian professor at Cambridge in 1701, but because of his increasingly unorthodox religious views, he was eventually driven from the university. In 1696 he published *A New Theory of the Earth* in which he traced the assumed periodic returns of this comet back to what he postulated as a prior collision with the earth that caused the biblical flood. To this collision he also ascribed the elliptical course of the earth around the sun and the rotational movement of the earth. His views, interestingly enough, were later echoed in part by Immanuel Velikovsky (1895-1979), a Russian-born psychoanalyst who in 1950 published *Worlds in Collision,* an extremely popular pseudo-scientific account of the interpretation of the Old Testament in which the planet Venus, in the guise of a comet, terrorized the earth.

Following the publication of his work, Halley continued to investigate the apparent periodic nature of the comet of 1682. His historical research convinced him that three earlier observed comets, those of the years 1456, 1380, and 1305, largely corresponded with the period and the elements assigned to the one of 1682. His identifications of comets sighted before 1456 later proved to be incorrect. But Halley's belief that they were previous returns of the same comet, together with his calculated elliptical orbits for the apparitions of 1607 and 1682, reassured him of the accuracy of his calculations and the certainty of his prediction.

The *Synopsis* was later reprinted in Halley's *Tabulae Astronomicae (Astronomical Tables),* the second edition of which contained additions to the original work, including general tables for motion of a comet in a very long ellipse. In it, Halley also firmly reiterated that the comet of 1682 would return again in late 1758 and advised the next generation of astronomers to watch for it. He concluded with the following appeal, "Wherefore, if according to what we have already said, it should return again about the year 1758, candid posterity will not refuse to acknowledge that this was first discovered by an Englishman."

Halley's prediction was of a new kind, not one about the alleged dire consequences of a comet's appearance, but about the appearance itself. He was the first to correctly predict the return of a comet, and thus document comet periodicity. And this discovery was a direct result of a form of cooperation between Newton and Halley, a bold partnership of subtlety and substance, of brilliance and practicality—and of benefit to both.

It is not known who first fixed Halley's name on this returning apparition. It is one of the few comets not named for its actual discoverer. But Halley's correct prediction marked an epoch in the history of astronomy, and posterity has not denied him the appropriate honor.

PARABOLIC ELEMENTS OF 24 COMETS (After Halley)*

Perihelion Passage (London Time)			Longitude of Ascending Node			Inclination			Motion	Longitude of Perihelion			Perihelion Distance (AU)
			°	'	"	°	'	"		°	'	"	
1337	2 Jun	6:25	84	21	0	32	11	0	R	37	59	0	0.40666
1472	28 Feb	22:23	281	46	20	5	20	0	R	45	33	30	0.54273
1531	*24 Aug*	*21:18*	*49*	*25*	*0*	*17*	*56*	*0*	*R*	*301*	*39*	*0*	*0.56700*
1532	19 Oct	22:12	80	27	0	32	36	0	D	111	7	0	0.50910
1556	21 Apr	20:03	175	42	0	32	6	30	D	278	50	0	0.46390
1577	26 Oct	18:45	25	52	0	74	32	45	R	129	22	0	0.18342
1580	28 Nov	15:00	18	57	20	64	40	0	D	109	5	50	0.59628
1585	27 Sep	19:20	37	42	30	6	4	0	D	8	51	0	1.09358
1590	29 Jan	3:45	165	30	40	29	40	40	R	216	54	30	0.57661
1596	31 Jul	19:55	312	12	30	55	12	0	R	228	16	0	0.51293
1607	*16 Oct*	*3:50*	*50*	*21*	*0*	*17*	*2*	*0*	*R*	*302*	*16*	*0*	*0.58680*
1618	29 Oct	12:23	76	1	0	37	34	0	D	2	14	0	0.37975

* Modified from the original in which the longitudes are measured by reference to the constellations of the zodiac and the perihelion distance is expressed in logarithms.

PARABOLIC ELEMENTS OF 24 COMETS (Continued)

Perihelion Passage (London Time)			Longitude of Ascending Node			Inclination			Motion	Longitude of Perihelion			Perihelion Distance (AU)
			°	'	"	°	'	"		°	'	"	
1652	2 Nov	15:40	88	10	0	79	28	0	D	28	18	40	0.84750
1661	16 Jan	23:41	82	30	30	32	35	50	D	115	58	40	0.44851
1664	24 Nov	11:52	81	14	0	21	18	30	R	130	41	25	1.02575
1665	14 Apr	5:15	228	2	0	76	5	0	R	71	54	30	0.10649
1672	20 Feb	8:37	297	30	30	83	22	10	D	46	59	30	0.69739
1677	26 Apr	0:37	236	49	10	79	3	15	R	137	37	5	0.28059
1680	8 Dec	0:06	272	2	0	60	56	0	D	262	39	30	0.00612
1682	*4 Sep*	*7:39*	*51*	*16*	*30*	*17*	*56*	*0*	*R*	*302*	*52*	*45*	*0.58328*
1683	3 Jul	2:50	173	23	0	83	11	0	R	85	29	30	0.56020
1684	29 May	10:16	268	15	0	65	48	40	D	238	52	0	0.96015
1686	6 Sep	14:33	350	34	40	31	21	40	D	77	0	30	0.32500
1698	8 Oct	16:57	267	44	15	11	46	0	R	270	51	15	0.69129

Table 3-1

CHAPTER 4

The Comet Returns

During six months we calculated from morning until night. . . .

—Joseph Jérôme Lalande

The event which Halley foretold elicited no great enthusiasm at the time. The imperfect state of mathematics, particularly celestial mechanics, prevented him from providing to the world proof of the certainty of his prediction. Since no way had yet been developed by which the effects of planetary gravitation on a comet's orbit could be calculated, any prediction had to be approximate. For many of his contemporaries all this remained purely conjecture. As the French astronomer Philippe Gustave de Pontécoulant (1795-1874) later wrote, "He could only announce these felicitous conceptions of a sagacious mind as mere intuitive perceptions which must be received as uncertain by the world, however he might have felt them himself, until they could be verified by the process of a rigorous analysis."

Nearly all of *celestial mechanics,* that is, that branch of astronomy concerned with the motion and position of heavenly bodies, including comets, is derived from Newton's laws of motion. The history of science is replete with examples of how slowly new ideas are accepted, and the *Principia* was no exception. Although Newton's reputation was great throughout

Europe and his book was considered remarkable by those few who could read it intelligently, its doctrines and its spirit were met with neither immediate nor universal acceptance. Christiaan Huygens (1629-1695), a famous Dutch physicist and astronomer best known as the founder of the wave theory of light, wrote in a letter of 1690 to his former student, the German mathematician and philosopher, Gottfried Wilhelm Leibniz: "Newton's principle of attraction seems to me absurd." He and other of Newton's scientific contemporaries found it difficult to accept a mathematical concept of a universal force with no explanation of how it works.

Huygens was an advocate of a competing system of "philosophy," that is, physical science, expounded by René Descartes (1596-1650) to account for the facts of astronomy as well as for such phenomena as gravity and magnetism. This was the theory of "vortices." It held that the originally undifferentiated state of the universe was disrupted by the local development of vortical or swirling motions. This theory dominated physics well into the eighteenth century. A partial reason for its success was that it allowed escape from Copernican theory and at the same time satisfied Catholic orthodox teachings. It was popular at the court of Louis XIV and for a time was even taught at Cambridge.

Newton had stated his laws verbally in the *Principia*. Ultimate acceptance of his theory depended upon the development of mathematical formulae describing the motion of a celestial body which satisfactorily agreed with accurate observations. This task was assumed by several eighteenth-century mathematicians whose faith in the Newtonian system gave them the zeal needed for such labors. Foremost among them were the Swiss-born mathematician Leonhard Euler (1707-1783) and his French-born contemporaries Jean Le Rond d'Alembert (1717-1783) and Alexis Claude Clairaut (1713-1765).

In 1736, Euler presented Newtonian dynamics in analytic form for the first time. He also devoted considerable attention, together with d'Alembert and Clairaut, to developing a more perfect theory of lunar motion, a particularly troublesome problem, since it involved the interactions of sun, moon, and earth—the so-called three-body problem.

Newton's law of universal gravitation states that all objects attract each other with a force directly proportional to the product of their masses and inversely proportional to the square of their separation. The problem of determining the motion of two bodies that are attracted to each other is called the problem of two bodies. This had been solved completely, and the solution goes by the name of Keplerian motion, but the solution had only limited application in practice.

The law of universal gravitation, however, is valid even when computing the interactions of any number of bodies. In particular, when three bodies are involved, there are deviations (called perturbations) from the elliptical orbits of Kepler. This is the celebrated three-body problem of celestial mechanics: What happens when three bodies interact gravitationally with each other? This problem, despite manned spacecraft to the moon and precision robot flights to the outer planets, is still unsolved today, except in special restricted circumstances.

During the late 1740s, Euler, d'Alembert, and Clairaut were all working, each of them with varying success, on the three-body problem with reference to the moon, sun, and the earth. The observed motions of the moon differed from the calculations based on Newton's theory of gravitation. In 1745, the calculations of both Clairaut and d'Alembert gave the value of eighteen years for the period of revolution of the *lunar apogee,* that is, the greatest distance of the moon from the earth, whereas observations showed this value to be nine years. This caused doubts about the validity of Newton's system and delighted the latter-day Cartesians who saw Newton's theory on the defensive. Clairaut suggested that Newton's law of attraction was incomplete and needed the addition of a correction factor. His rival d'Alembert disputed this modification, and Clairaut himself subsequently realized that the error lay not in the law, but in the calculations.

Finally, in 1749, Clairaut announced that the difference between theory and observation was due to the fact that he and others solving the corresponding differential equation had not completed the solution. When he extended his calculations, he found that, in accordance with Newton's theory, the apogee of the moon moved over a time period very close to that noted by observations. This was the first approximate resolution of the three-body problem in celestial mechanics.

Shortly thereafter, Euler developed his own solution to the three-body problem, the so-called first Euler lunar theory. He also described another analytic method known to mathematicians as a variation of elements technique. Euler's calculations also confirmed Newton's theory of gravitation.

But Clairaut had beat out his rivals and, strengthened by this success and encouraged by his friend, Joseph Jérôme Lalande (1732-1807), he turned his remarkable skills to the movement of comets, a special case of the three-body problem.

Clairaut was ably suited for this task, since he was one of the finest mathematicians of his generation (Fig. 4-1). Born in Paris in 1713, the son of a mathematics teacher, he received no formal education, but was instead tutored by his father. Clairaut was a precocious child and at the early age of eighteen was elected to the French Academy of Sciences. In 1736, he journeyed to Lapland where he helped determine the length of the meridian and confirmed the Newtonian view that the earth is flattened at the poles. A formula was named after him, expressing the earth's gravity as a function of latitude. He authored several works on celestial mechanics, did research on the calculus, and assisted in the French translation of the *Principia*. Clairaut was a vivacious, handsome man who was

Fig. 4-1. Alexis Claude Clairaut (1713-1765). He calculated the perihelion passage of Halley's comet in 1759 to within a month, confirming Halley's prediction and validating Newton's theory of gravitation. Although proposed for election to the Académie des Sciences in 1729 at the age of sixteen, Clairaut was not confirmed by the king until two years later.

successful with women, yet remained unmarried. Following a brief illness, he died at the age of fifty-two.

As 1758 drew near, astronomers naturally recalled Halley's prediction and Clairaut found in the comet's anticipated passage a new field of activity. Halley's prediction was based mathematically solely on the gravitational attraction of the sun, with no consideration of the perturbations caused by the other planets. It is true that he suggested that the influence of Jupiter would retard the return of the comet, but he could not prove this. It was now possible, as it had not been seventy-five years earlier, to determine mathematically the effect of planetary perturbations. It was also possible to convert Halley's conjecture into a precise astronomical prediction, since the disturbing effects of both Jupiter and Saturn could be taken into account.

It was now 1757 and Clairaut began a race against time, attempting to calculate as accurately as possible the date of the comet's perihelion passage. But the conditions involving a comet, a planet, and the sun were not as favorable for calculation as those in the earlier three-body problem. The great eccentricity of the comet's orbit made it necessary to attack the problem in a very different manner from that employed in the case of the moon, the planets, and the sun.

What had to be established was an *ephemeris* for Halley's comet, that is, a table listing its place in the firmament at stated intervals of time. This required the solution of two very different classes of problems. The first was the development of analytical formulae applicable to a body attracted by the sun and disturbed by a planet. For this, Clairaut used a modified version of his lunar theory by which to ascertain the extent of the perturbations of the comet's orbit likely as a result of the influence of Jupiter and Saturn, which might either hasten or delay its movement.

Having devised the methods necessary for this novel task, Clairaut then confronted the second half of his problem: the arithmetical calculations. Clairaut was a theorist, not a man of a practical bent. For the task of the prodigious numerical calcu-

Fig. 4-2. Joseph Jérôme Lalande (1732-1807). He assisted Clairaut in the calculations involved in predicting the perihelion date for the return of Halley's comet in 1759. The task was so enormous that his health was permanently affected. Lalande was extremely well known during his lifetime because of his popular writings.

lations involved, he enlisted the assistance of Joseph Jérôme Lalande (Fig. 4-2) and of Nicole Reine Étable de Labrière Lepaute, the wife of an eminent Parisian watchmaker.

The name of Madame Lepaute was omitted in Clairaut's later account of the computation, possibly according to Lalande, for fear of offending another lady of whom Clairaut was enamored. Her labors, assisting both Clairaut and Lalande in the work of calculation, at length so weakened her sight that she was forced to stop. She died in 1788 while nursing her invalid husband. Her name deserves more than a note in the annals of astronomical science.

The magnitude of these mathematical calculations can hardly be exaggerated. It was necessary first to compute the distances of the comet from the planets Jupiter and Saturn since 1682, when it was last observed, and then those for the previous revolution as well—an interval of more than 150 years. This alone would have been laborious enough; but the corollary part of the work, in which the disturbing force of each planet during this long period had to be taken into account, involved much greater and more intricate calculations. The disturbing action can be estimated for only one planet at a time, and therefore it becomes necessary to repeat all the numerical computations for each disturbing body. Moreover, the disturbing action of a planet can be computed for only an extremely brief segment of the comet's orbit at a time. It is the sum of all these that comprise the whole effect.

Lalande described the task they accomplished. "During six months we calculated from morning until night, sometimes even at meals, the consequence of which was that I contracted an illness which changed my constitution for the remainder of my life. The assistance rendered by Madame Lepaute was such that, without her, we never could have dared to undertake this enormous labor, in which it was necessary to calculate the distance of each of the two planets, Jupiter and Saturn, from the comet, and their attraction upon that body, separately for every successive degree, and for 150 years."

Their computations were brought to a close early in November 1758. As a result of their work, it was determined that the comet would be delayed for 518 days by the attraction of Jupiter and for 100 days by that of Saturn. Clairaut, fearing that the comet might anticipate his announcement, stated in a memoir presented at the opening session of the Academy of Sciences in Paris on 14 November 1758 that passage to perihelion would occur about 15 April 1759. He did this, however, with reservations.

In the course of his presentation, he remarked on the interest and curiosity the general public was showing in the matter. He was sure that all true lovers of science desired the return of the comet, since it would afford striking confirmation of Newton's theory. He mentioned others who were hopeful that it would not return, and so prove the reasoning of astronomers to be on a level with the fanciful theories put forward by dreamers and speculators. Against such opponents of science, Clairaut declared himself wholeheartedly in favor of the law of universal gravitation. But he had been pressed for time in the calculations, and so Clairaut added, "You can see the caution with which I make such an announcement, since so many small quantities that must be neglected in methods of approximation can change the time by a month."

The allowance for an error of one month was the result of the unfavorable circumstances under which the computations were made. Certain factors had had to be omitted and there was a possibility that errors in calculation had been made. Clairaut had been forced to determine the elements of the orbit based on the observations of Peter Apian during the comet's return in 1531. These, however, were far from accurate, having been made at a time when only casual attention was paid to comets, and with no awareness of their possible future importance. The masses of the planets, Jupiter and Saturn, which Clairaut had adopted were later found to be in error—especially that of Saturn. Pierre Laplace (1749-1827), the French astronomer and mathematician, remarked that had the true mass of Saturn been known, the error could have been reduced to thirteen days. Clairaut neglected the disturbing action of the earth, which was not inconsiderable, as well as the other inner planets, Mars, Venus, and Mercury. He was not aware of the existence of Uranus, Neptune, and Pluto, and thus their effect was not computed.

Clairaut, however, knew the potential of error in his calculations. In an astute observation before the appearance of the comet, he said, "A body which passes into regions so remote, and which is absent from our observation for intervals so protracted, may be submitted to forces altogether unknown to us; such as the attraction of other comets, or even of some planet too distant from the sun ever to be seen from the earth." A quarter century later in 1781, the planet Uranus was discovered by Sir William Herschel.

Newton and many of his followers also believed that there was an ethereal fluid which filled all of space and acted as a "resisting medium." This would, they assumed, have the effect of decreasing the period of the comet upon successive returns by diminishing the velocity of its centrifugal force and thus increasing the solar attraction. No one, however, had attempted to quantify such an effect until Clairaut, who endeavored to determine its influence upon the motions of the comet. He calculated that the acceleration would not exceed seven and one-half minutes.

After presenting his data to the Academy, Clairaut returned to his calculations, revising them and completing some which he had not time to execute previously. Several weeks later, he presented to the Academy his prediction of 4 April 1759 as the comet's arrival at the nearest distance to the sun.

The actual time of perihelion passage in 1759 was 13 March. Clairaut's initial prediction of 15 April had missed by only two days his predicted one month limit of error. Some historians, however, have incorrectly stated his forecast as 13 April to give him the benefit of a more accurate prediction.

Later, by using a different method, he arrived at 31 March 1759 as the date of passage, and for this paper he was awarded a prize by the Academy of St. Petersburg in 1762. Clairaut had proved that such was the strength of the Newtonian mechanics that a man could calculate within one month a comet's passage over a period of 150 years.

When Clairaut first announced his results, his nemesis d'Alembert questioned both his calculations and his method. To this Clairaut replied, "If one wishes to establish that Messrs. d'Alembert et al. can solve the three-body problem in the case of comets as well as I did, I would be delighted to let him do so. But this is a problem that has not been solved before, in either theory or practice."

Lalande, however, felt certain of the calculations and even more certain of the comet's return. But he was less confident that it would be sighted by the astronomers. He had noted that in its successive appearances subsequent to 1456, the comet had gradually decreased in both magnitude and splendor. The comet of 1456 occupied a space nearly 70 degrees in length. Its appearance had spread terror throughout Europe. In 1607, its appearance was witnessed by both Kepler and Longomontanus. The comet resembled a star of the first magnitude with a barely discernible tail. The most recent return in 1682 had excited little attention except among astronomers. Lalande assumed this decrease of magnitude and brilliancy to be progressive and that on its expected return the comet might escape the observation even of astronomers. He did not consider it possible that such a great change in physical appearance might simply be caused by the comet's position with respect to the earth and sun at the time of its visibility. "We cannot doubt," he said, "that it will return; and even if astronomers cannot see it, they will not therefore be the less convinced of its presence; they know that the faintness of its light, its great distance, and perhaps even bad weather, may keep it from our view; but the world will find it difficult to believe us; they will place this discovery, which has done so much honor to modern philosophy, among the number of chance predictions. We shall see discussions spring up again in the colleges, contempt among the ignorant, terror among the people, and seventy-six years will roll away before there will be another opportunity of removing all doubt."

As 1759 approached, scientific interest rose to fever pitch. Voltaire (1694-1778), the famous French writer, remarked that the astronomers of France did not sleep for fear of missing the long-awaited comet. The most elaborate attack was prepared by Joseph Nicolas Delisle (1688-1768).

Delisle was a distinguished astronomer of the day and numbered Lalande among his students. In 1718, he was appointed to the chair of mathematics at the Collège Royal. At the request of Peter the Great he had spent many years in Russia, founding both an observatory and school of astronomy there. In 1755, he presented his large collection of geographical and astronomical material to the French government and received in return a life annuity of 3,000 livres. His assistant in these later years was a remarkable individual named Charles Messier (1730-1817). At his observatory in the Hôtel de Cluny in Paris, Delisle set Messier (Fig. 4-3) the task of systematically searching for the anticipated comet.

Messier arrived in Paris at the age of twenty-one with neat handwriting and some drawing experience as his sole recommendations. But he was hired by Delisle, and soon thereafter he began a life devoted to the search for comets. He received from Louis XV the appellation of *le furet de comètes (the ferret of the comets)*. During his lifetime Messier discovered twenty-one comets, none of them however of any particular interest. To aid him in his searches, he compiled a catalogue of all hazy objects in the sky which might be confused with comets just beginning to shine. This catalogue

Fig. 4-3. Charles Messier (1730-1817). He is shown in this portrait at the age of 40. Despite a search which lasted for eighteen months, this noted comet watcher was frustrated in his quest to first observe the predicted return of Halley's comet by a peasant farmer.

of nebulae and clusters remains his most enduring contribution. M31 or Messier Object 31 is today recognized as the Andromeda Galaxy.

Messier had no taste whatever for theoretical research, but he did have a genuine passion for observation. Jean François de La Harpe (1739-1803), a French poet and literary critic, writing in 1801, said that "he passed his life in search of comets.... He was an excellent man, but had the simplicity of a child. At a time when he was in expectation of discovering a comet, his wife took ill and died. While attending upon her, being withdrawn from his observatory, Montagne de Limoges anticipated him by discovering the comet. Messier was in despair. A friend visiting him began to offer some consolation for the recent affliction he had suffered, but Messier, thinking only of the comet, exclaimed, 'I had discovered twelve. Alas, that I should now be robbed of the thirteenth by Montagne!' and his eyes filled with tears. Then, remembering that it was necessary to mourn for his wife, whose remains were still in the house, he exclaimed, 'Ah! that poor woman,' and again wept for his comet."

In 1757, Delisle computed an ephemeris and prepared charts that plotted the anticipated course of the awaited comet. He gave this to Messier to guide him in his search. The area to be surveyed, however, proved to be too restrictive and Messier's attention was directed to an area of the firmament through which the comet did not pass. Night after night for eighteen months Messier observed. Finally, despite the erroneous charts and after several days of cloudy weather, he succeeded on 21 January 1759 in finding the elusive comet. Delisle, however, who seemed to think that he had a property right in Messier's observations, demanded strict secrecy. Jean-Baptiste Joseph Delambre (1749-1822), a historian of eighteenth-century astronomy, recorded that he hoarded this discovery as a miser does his wealth, neither using it himself nor allowing anyone else to do so.

Unknown to these French astronomers, Halley's comet had already been observed nearly a month before by a farmer in Saxony by the name of Johann Georg Palitzch. Only after this news reached Paris did Delisle reveal Messier's discovery.

Palitzch, living at Problis, a town near Dresden, first saw the comet on 25 December 1758. While observing the variable star Omicron Ceti with his homemade reflecting telescope of eight-foot focal length, he found between Delta and Epsilon Ceti a nebulous body which proved to be the comet. His discovery was not accidental, for Palitzch knew about Halley's prediction and had been watching for the comet.

Sir John Herschel called him "a peasant by station, an astronomer by nature." He worked his farm by daytime; in the evenings he read books on science and mathematics. Baron Francis von Zach (1754-1832), a German astronomer of repute, was personally acquainted with Palitzch and knew him to be a careful and constant observer of the heavens. He had good vision, and was in the habit of scanning the heavens with the naked eye. This may have given rise to the notion that he spotted Halley's comet with the naked eye at the same time professional astronomers were fruitlessly searching for it with telescopes.

The next day Palitzch communicated his discovery to a Dr. Hoffman, who saw the comet on the evenings of the 27th and 28th of December. A few days later the comet was discovered independently by an anonymous astronomer at Leipzig. "But," says Count de Pontécoulant, "jealous of his discovery, as a lover of his mistress, or a miser of his treasure, he would not share it, and gave himself up to the solitary pleasure of following the body in its course from day to day, while his contemporaries throughout Europe were vainly directing their anxious search after it to other quarters of the heavens."

In Paris, since the night of his initial sighting, Messier had continued to observe the comet until 4 February. It was positioned low in the southern skies near the horizon and was barely visible because of the effect of twilight. Consequently, little astronomical investigation of the comet could be done. At this time its shape was very round and it had a brilliant nucleus well differentiated from the surrounding nebulosity. No tail, however, was seen. Toward the end of the month the comet was lost in the sun's glare as it approached perihelion. Messier was the only astronomer who professed to have seen the comet just before it became lost in the sun's rays. Its reappearance from behind the sun was also noted by Messier before sunrise on the morning of 1 April. He stated that he was able to distinguish the tail through his telescope.

By 17 April, the comet ceased to be visible in the morning, when because of its southern position it rose after the sun. But as its position continued to change, it again appeared after sunset on 29 April, and Messier described it as having the appearance of a star of the first magnitude. Unfortunately, by this time the light of the moon was so strong that it diminished the effect of the comet.

During the months of April and May, nearly all the observatories in Europe sighted the comet, but it was seen to best advantage in the Southern Hemisphere, although records of its appearance there are scanty. It was observed at Pondicherry in India by Father Coeurdoux and at the Isle de Bourbon, now called Réunion, in the Indian Ocean, by the French astronomer Nicolas Louis de La Caille (1713-1762). Both astronomers noted that the tail was distinctly visible to the naked eye and that on 5 May its length measured forty-seven degrees.

Messier's extended observations were the most accurate of the 1759 return. Though the honor of being the first to welcome the wanderer was denied to him, he had the satisfaction on 3 June of being the last to see it off on its long, solitary journey to a region far beyond the boundary of the inner planets of our solar system.

The return of the comet and the success of the computations concerning its return were rightly regarded as one of the most elegant triumphs of Newton's theory of gravitation, putting an end forever to the theory of Cartesian vortices. Halley's prediction, made more than fifty years before, had been triumphantly vindicated. It was fitting that the comet now bore his name.

CHAPTER 5

The Later Apparitions

At least two days of this error must be attributed to causes other than errors of calculation or errors in the adopted positions and masses of the planets.

—Cowell and Crommelin

During the three-quarters of a century that passed before the comet's next appearance, great telescopes were constructed. Especially notable were those at Dorpat (now Tartu) in Estonia and at the Cape of Good Hope in South Africa. In 1781, the planet Uranus was discovered by Sir William Herschel (1738-1822), who thus added yet another body whose disturbing effect upon the comet would have to be taken into account. The mass of Saturn was more accurately determined, and other great advances in the field of mathematical astronomy were made. Societies were established which sought to advance knowledge by offering prizes to those responsible for solving vexing scientific problems. In 1778, the French Academy of Sciences offered an award for work on the determination of cometary orbits and the perturbations which these orbits experience by the action of the planets near which the comets pass.

Joseph Louis Lagrange (1736-1813) was a Frenchman born in Italy who is considered by many to have been the greatest mathematician of the eighteenth century. His interest in mathematics had been aroused by the chance reading of an essay by Edmond Halley. In 1783, he presented a paper for computing the perturbed motion of a comet using a variation of elements technique. His general method served as the basis for the determination of the next return of what was then the only known periodic comet, Comet Halley. For his efforts Lagrange was awarded the prize by the French Academy.

As the time of the comet's reappearance drew near, both public and professional interest grew. No less than five independent computations of the orbit of the comet were made, and it was expected that a high degree of exactness in predicting the time of perihelion would result. Two French astronomers, Count de Pontécoulant and Baron Damoiseau, two German mathematicians, Dr. Lehmann and Professor Rosenberger, and a lone Englishman, John Lubbock, competed for national honor.

According to an article in the *Astronomical Journal* of June 1977 by Donald K. Yeomans, their methods differed "only in how many perturbing planets were included, how many orbital elements were allowed to vary, and how many times per revolution the reference ellipse was rectified by adding the perturbations in elements." All of them adopted similar estimates of the masses of the disturbing planets, including that of Uranus, although its mass was imperfectly known.

The Academy of Sciences at Turin offered a prize for an essay on the perturbations undergone by Halley's comet since 1759. This competition was opened to astronomers of all nations, but Baron Marie Charles Damoiseau of Paris was the successful candidate. The details of his calculations were published in the twenty-fourth volume of the *Memoirs of the Academy of Turin* for 1817. His initial prediction of 16 November 1835 as the time of perihelion passage was remarkably accurate—this was in fact the actual date of the perihelion. However, several years later he added the perturbations due to the earth to his calculations and revised his prediction for 4 November 1835.

In 1826, the Academy of Sciences of Paris proposed a similar prize, having twice before offered it without anyone coming forth to claim it. On this occasion Count Philippe Gustave de Pontécoulant aspired to the honor. "After calculations," he recalled, "of which those alone who have engaged in such researches can estimate the extent and appreciate the fastidious monotony, I arrived at a result which satisfied all the conditions proposed by the Institute. I determined the perturbations of Halley's comet by taking into account the simultaneous actions of Jupiter, Saturn, Uranus, and the Earth; the comet having passed in 1759 sufficiently near our planet to produce in it (the comet) sensible disturbances; and I then fixed its return to its nearest point to the sun for the 7th of November, 1835." Subsequent to this prediction he made additional calculations, taking into account a newly determined value for the mass of Jupiter, and concluded that the perihelion passage would take place on 14 November. In 1829, he was awarded the prize by the French Academy.

The investigations and subsequent predictions by both Damoiseau and Pontécoulant could never have been truly accurate, since they omitted the gravitational influence of several of the planets. Nor had they given sufficient weight to the actual orbit of the comet in 1759. Since 1759 was the starting point from which to determine the orbit for 1835, it was important to obtain the most accurate knowledge possible of the conditions of its 1759 passage.

The most elaborate calculations of the 1835 return were undertaken by Professor Otto August Rosenberger (1800-1890) of the University of Halle in Germany. He felt that he should go back much further than either Damoiseau or Pontécoulant had done. Accordingly, he recomputed the orbit for both the 1759 and 1682 apparitions. Then for the interval between 1682 and 1759, Rosenberger computed the effect of the seven known planets upon the comet. Next he computed the perturbations for the 1759 to 1835 interval. In addition, he calculated the effect of the supposed "resisting medium" which pervaded the space between the planets. Such a substance had been invoked in 1823 by Johann Franz Encke (1791-1865) to account for the anomalous motion of a comet initially discovered in 1786. Such a resistance would act to speed the comet's return by diminishing its centrifugal force, thus reducing its orbit and shortening its period. Rosenberger's results showed that the smaller planets, Earth, Venus, Mercury, and Mars, would together hasten the return by a total of twenty-two days, and that perihelion would occur on 12 November 1835 if the "resisting medium" was not taken into account and on 4 November if it were.

So complete were Rosenberger's computations that he was generally accorded the honor of having conducted the best and most elaborate investigation. The Royal Astronomical Society of Great Britain awarded him their gold medal in 1837 for his efforts. Although his perihelion prediction was four days too early, his gravitational computations were quite accurate. He also published an ephemeris of the comet in which its exact route in the heavens was designated. During the presentation of the medal, George Biddell Airy, President of the Society, commented that so complete were his calculations "...that if names were taken, not from the discoverers of these bodies, or from those who conjecture their identity, but from those who, by accurate calculations on a uniform system, combine the whole of our information relating to them, we should call this body, not Halley's, but Rosenberger's Comet."

Rosenberger, however, had a competitor in his own country—Dr. Jacob Lehmann, who felt that there was room for another investigation of the elements and disturbances of the orbit of Halley's comet. His computational method was similar to that of his countryman, except that he took the comet's passage in 1607 as his starting point. On 25 July 1835, just prior to the initial sighting of the comet, he announced the date of perihelion would be 26 November 1835. Because he had paid insufficient attention to slight changes in the orbit which had occurred since the comet's previous appearance, his prediction was ten days later than the actual event. Consequently, he collected no prizes or medals for his efforts.

The English astronomer, Sir John William Lubbock (1803-1865), who also ascertained the true orbit of the comet in 1759, had the most inexact prediction: 31 October 1835.

In a paper published late in 1834, a German astronomer in Bremen, Heinrich Olbers (1758-1840), had suggested that the comet might be discovered much sooner than was expected. His examination of the comet's previous appearances and his investigation into the brightness of other comets, particularly the comet that appeared in 1811, convinced him that an early sighting was possible. He said that the path of the comet between December and April would lie in the constellations of Auriga and Taurus. This part of the sky was favorably situated for the observations in Northern and Central Europe and diligent watch was kept over this region throughout the winter months of 1834-35.

Sir John Herschel (1792-1871), the son of Sir William Herschel, had journeyed to the Southern Hemisphere in 1833 to continue the systematic mapping of the stars begun by Edmond Halley. Employing his great reflecting telescope with the twenty-foot focal length at the Cape of Good Hope, he meticulously swept the night sky. But to no avail; the comet was not sighted. The Auriga-Taurus region was lost to view during *conjunction* with the sun in the summer months, and the astronomers had to wait patiently for its reappearance.

Finally, on the morning of 6 August, near the star Zeta Tauri, a faint misty object, barely discernible under clear Italian skies even with the aid of a powerful telescope, was observed by the Frenchman, Father Dumouchel (1773-1840), Director of the Observatory at the Collegio Romano. The comet was located close to the spot Rosenberger had predicted for that day. Unfavorable weather and moonlight during the next few days delayed its discovery elsewhere.

On 21 August, it was seen again by Friedrich Georg Wilhelm von Struve (1793-1864) at Dorpat, and by Professor Johann von Lamont (1805-1879) at Munich. The Dorpat observations showed that the error of Rosenberger's predicted placement was only seven minutes of arc in right ascension and seventeen minutes in declination. The first observation in the United States was by Professor Loomis at Yale College, who sighted the comet on 31 August. Struve first saw it with the naked eye on 23 September and the first beginnings of a tail were noted shortly thereafter.

During the succeeding month the comet rapidly increased in brightness and during its course through the constellations Ursa Major, Hercules, and Ophiuchus, it put forth a train of ever-increasing length. By mid-October the tail had reached a maximum length of between 20 degrees and 30 degrees, depending upon the estimation of the observer. Thereafter, the comet became less brilliant and the tail gradually shortened, vanishing about the time of perihelion as the comet sank below the southwestern horizon.

Hidden by the sun from 22 November until 30 December, the comet passed over to the Southern Hemisphere, becoming visible at the Cape of Good Hope on 24 January 1836, where it was observed by Sir John Herschel and Thomas Maclear (1794-1879) under very favorable conditions until 5 May. The last glimpse of the comet was made by Lamont on 17 May at Munich. It had been observed for a total of 286 days during this return.

This apparition of Halley's comet was a particularly bright one in both hemispheres. It had been the expectation of astronomers that at this return new data might be obtained from which a satisfactory theory regarding the physical nature of cometary bodies might be formed. No sooner had the comet become generally visible than extraordinary changes were observed in its physical appearance. Since there was as yet no way to document these phenomena with photographs, vivid description and sketches are the only record.

The comet's nucleus was variously described as a "fan-shaped flame," a "red-hot coal of oblong form," a "blazing rocket," and a "powder horn." The halo or coma was noted to diminish or increase in size, presenting the nucleus sometimes as a sharp, well-defined body like a faint star and sometimes as a nebulous blotch. But the nature of the comet's substance remained a mystery. With this apparition, however, the elements of Halley's orbit were more accurately defined, its motion was confirmed as being in a direction contrary to that of the planetary system, or *retrograde,* and its course far removed from the ecliptic in the more remote parts of the orbit.

Before the next apparition, the science of astronomy underwent considerable advancement. In 1846, the giant planet Neptune was discovered. The masses of all the known planets were determined to a greater accuracy than ever before. New methods of computation were devised. The conditions were thus extremely favorable for a precise prediction of the comet's next perihelion.

It was surprising, therefore, that the initial two independent calculations for the predicted return showed a wide discrepancy. In 1862, Dr. Anders Jonas Ångström (1814-1874), the man for whom the angstrom unit of light measurement is named, researched twenty-five previous passages of the comet. He calculated that the average period was 76.93 years. He derived a formula and plotted a curve by which he could determine the time of perihelion to within one year for all the previous returns, and in most cases to within less than half a year. According to him, therefore, the next passage should occur early in 1913. At about the same time, Count de Pontécoulant, encouraged by his success with the 1835 return, set about the task of calculating the next passage. He took into account the perturbative effects of

Jupiter, Saturn, and Uranus, and predicted in 1864 in the *Comptes Rendus de l'Académie des Sciences* that the date would be 24 May 1910. There was a discordance between the two predictions of 2.7 years. Some concern was noted regarding the accuracy of Pontécoulant's calculations, inasmuch as the disturbing action of only three planets was considered and there were some obvious arithmetical errors in his paper. It was also felt that he may well have given less attention to these calculations than to those for 1835.

But as the next return rapidly approached, no further investigations had been made. In a report published in the *Monthly Notices of the Royal Astronomical Society* for 14 December 1906, A.C.D. Crommelin wrote, "One can hardly imagine a greater loss of prestige to astronomy than that which would arise if there were a notable error in the prediction of this return of the comet, after the wonderful success achieved in 1759 and 1835." Appealing to his fellow astronomers, he sought a re-examination of the orbit so that a close prediction could be made. As things turned out, it was Crommelin and his associate Cowell who were the most accurate in predicting the return.

Andrew Claude de la Cherois Crommelin and Philip Herbert Cowell were both astronomers at the Royal Observatory at Greenwich. Assisted by several volunteer computers, they began a study of the motion of the comet under the perturbative influence of Venus, Earth, Jupiter, Saturn, Uranus, and Neptune. They used a direct integration method, known as "Cowell's method," which was based upon a rectangular coordinate system and in which the orbit was rectified at every step of integration. They started with the comet's known position in 1835, working forward to 1910, then backward to 1759. They eventually arrived at a predicted 1910 perihelion passage date of 16 April. For their work they were awarded the 1,000 deutsche mark Lindemann Prize of the Astronomische Gesellschaft and honorary D.Sc. degrees from Oxford

University. The completeness of their examination prompted the opinion that if the comet did not return at the time they said it would, then most probably some hitherto unknown disturbing cause was acting on the comet.

As 1910 approached, once again all the comet hunters eagerly scanned the heavens in hopes of being the first to spot the comet's reappearance. Only this time the photographic plate could be used, a resource not previously available. More sensitive than the human eye, the photographic study of comets was pioneered at the Lick Observatory, near San Jose, California, by Edward Emerson Barnard (1857-1923), perhaps the foremost observational astronomer of his time. In 1892, he made the first discovery of a comet photographically. From then on no one attempted to sketch a comet, since the photograph provided more richness and accuracy of structural detail. As early as 22 December 1908, the American astronomer, Oliver J. Lee, at the Yerkes Observatory at Williams Bay, Wisconsin, commenced a photographic search for Halley's comet.

As the search intensified, so did nationalist fervor. Each country hoped its astronomers would detect the comet first. In September 1908, Herbert H. Turner (1861-1930), Savilian Professor of Astronomy at the University of Oxford, stated, "...it could scarcely be called selfishness to indulge in the pious hope that the good fortune of first detecting the comet should fall to an Englishman."

But this was not to be. Maximilian Wolf (1863-1932) of the Königstuhl Observatory at Heidelberg (Fig. 5-1), an eminent astronomer of the time who was noted for his innovative photographic methods, exposed photographic plates on 28 August 1909 which showed the faint image of the comet. He announced his discovery on 11 September. At once other observers searched their early plates. At Greenwich Observatory two plates had been taken on 9 September, but because of moonlight the length of exposure was limited. At first the comet was not detected, but after re-examination based upon Wolf's

Fig. 5-1. Maximilian Wolf (1863-1932). He was the first person to photographically identify Halley's comet. He announced his discovery on 11 September 1909 when the comet was more than 300 million miles distant.

coordinates, a very faint image was found on each plate. Mr. Knox-Shaw of the Helwan Observatory in Egypt had unknowingly photographed it as early as 24 August with the Reynolds Reflector. The guiding telescope used with this reflector was the seven-foot focal length refractor that John Herschel had used when he had observed the comet at the Cape of Good Hope 73 years earlier.

The comet at this time was over 300 million miles from the sun and a little further from the earth. Its position differed from the computed ephemeris by six minutes of arc in right ascension and four minutes in declination, bringing the perihelion date to 19 April 1910. A month earlier than de Pontécoulant's date and slightly three days later than that calculated by Cowell and Crommelin, this return was the shortest on record, 74 years and 5 1/2 months, due mainly to the near maximum disturbing effect caused by its close approach to Jupiter during the 1835 passage.

The comet was first seen non-photographically by Professor Sherburne Wesley Burnham (1838-1921) on 15 September 1909, using the 40-inch refractor at Yerkes Observatory. Shortly thereafter, many others reported sighting it. During the remainder of the year, the comet gradually brightened. On 18 November, a faint curved tail was noted, and for the next few months sudden variations in brightness occurred. On 11 February, Max Wolf at Heidelberg saw the comet with the naked eye. During the latter part of March and early April, it passed out of view behind the sun. Upon its reappearance, variations in the shape of the tail were noted daily. During the early part of May, it became a magnificent object in the morning sky (Fig. 5-3).

On 8 May, there was a total eclipse of the sun which afforded a unique opportunity to visualize Halley's comet. As comets approach the sun they generally increase in size, but the glare of the sun prevents them from being seen. During the total phase of an eclipse that blinding glare is cut off, providing an unusually favorable opportunity for observation. During the eclipses of 1882 and 1893, hitherto undetected comets were

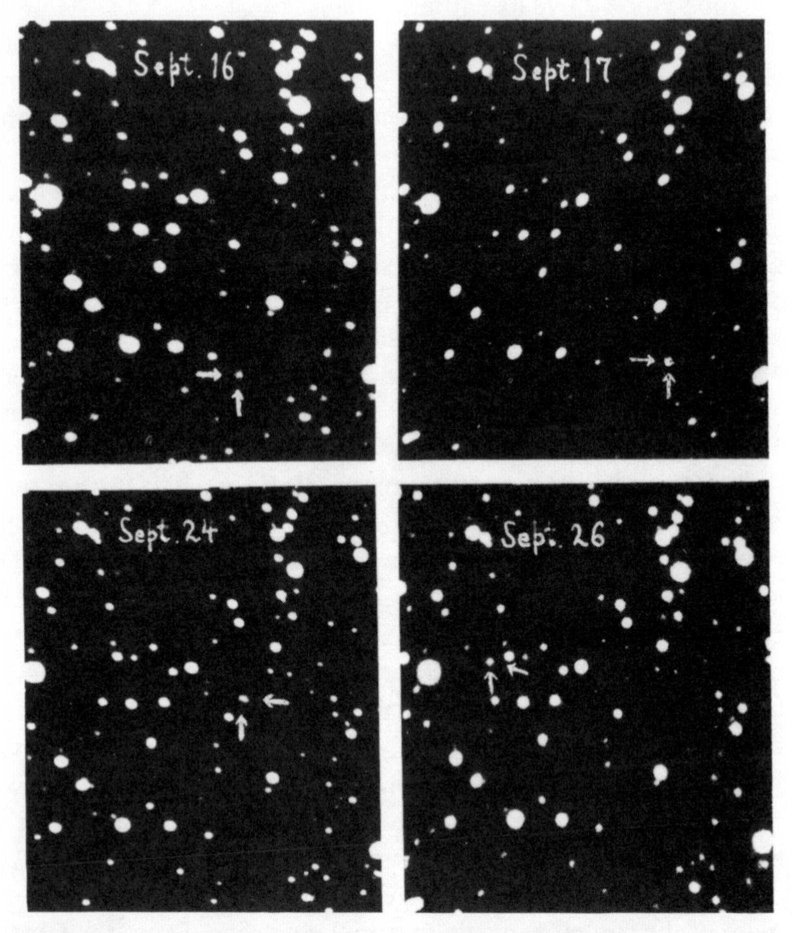

Fig. 5-2. First Photographs of Halley's Comet Taken in the United States. Oliver J. Lee of Yerkes Observatory used the two-foot reflector to obtain these four photographs of the same area of the sky during September 1909. The arrows indicate the position of the comet and also the components of its direction of motion. The comet is approximately magnitude 16.

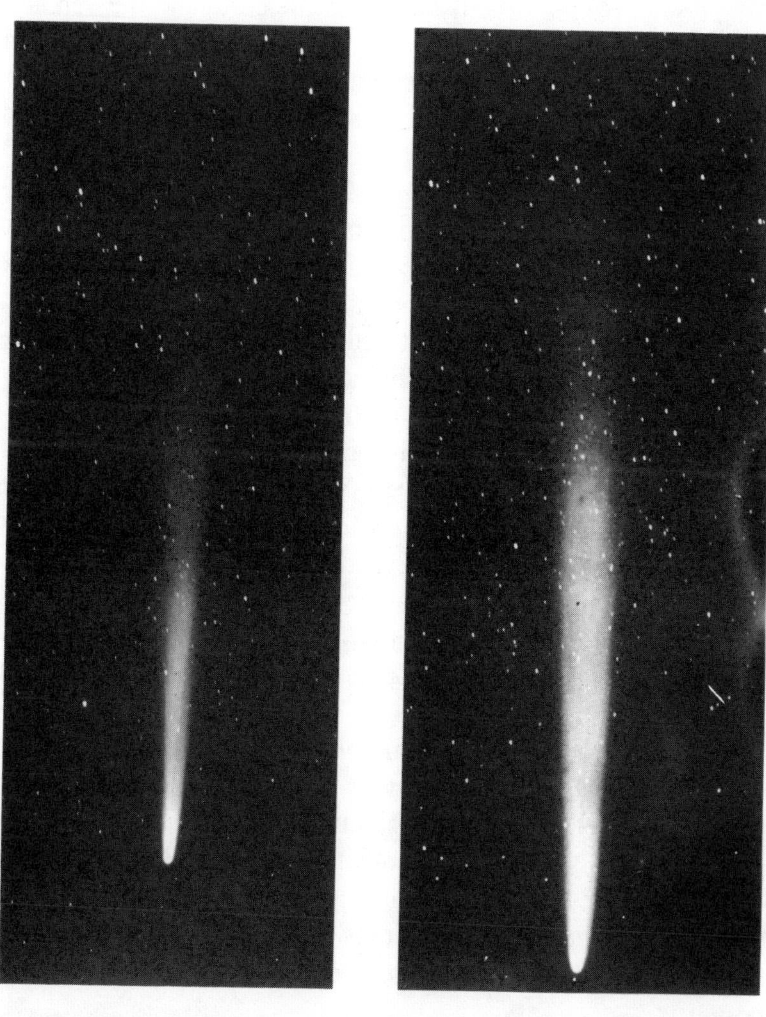

Fig. 5-3. Halley's Comet During May 1910. These two photographs were obtained in Honolulu and show the comet on 12 May (left) and 15 May with tail lengths of 30 degrees and 40 degrees respectively.

discovered near the sun. Unfortunately, however, with the exception of a small area of Tasmania, the track of this 1910 eclipse ran across the ocean, providing no suitable sites for observation.

Another unusual event occurred on 18 May, when the head of the comet passed directly between the earth and the sun at a distance from the earth of approximately 15 million miles. In both Europe and America this phenomenon occurred during the night. In order to witness the possible transit of the comet's head across the sun's disk, the American astronomer Ferdinand Ellerman (1869-1940) of the Mount Wilson Observatory mounted a scientific expedition to Hawaii. Observing the sun with a six-inch telescope, he could detect neither a trace of the nucleus of the comet against the face of the sun, nor any change in the brightness of the sun. It had been computed that a solid body as small as 200 miles in diameter would have been visible. This was confirmation of the theory that the density of comet material was very low despite its enormous volume.

Also on 18 May, the earth was expected to pass through the comet's tail. Nearly 200 meteorological stations in the United States and the West Indies were alerted by Professor Willis L. Moore, Chief of the U.S. Weather Bureau, to the possibility of auroral displays, meteoric trails, luminous clouds, solar and lunar halos, "...and all other appearances that may seem unusual and worth noting."

Unusual atmospheric phenomena were in fact reported by widely scattered trained observers at a time coinciding with the passage of the earth through the tail of the comet. Although the possibility that somehow the comet produced these effects could not be definitely proven, the display of iridescent clouds, caused by the presence of tiny particles within certain levels of the earth's atmosphere, strongly suggested an association with the comet.

In the southern portion of the Northern Hemisphere and in the Southern Hemisphere itself, the most brilliant object in the sky was Halley's comet with a tail stretching across the firmament for a distance of 120 degrees to 140 degrees and appearing as a broad band much like the Milky Way. This was the zenith of its brilliance, however, and as the comet receded from the sun it began to fade. It was last seen visually in a telescope at Helwan in Egypt on 29 April 1911 and was followed photographically until 1 July when it was 520 million miles from the sun and beyond the orbit of Jupiter.

Finding that their prediction for perihelion passage had been too early, Cowell and Crommelin re-worked their calculations. The discrepancy still persisted. "It now appears from the observations," they said, "that the predicted time is 3.03 days too early. At least two days of this error must be attributed to causes other than errors of calculation or errors in the adopted positions and masses of the planets." As with the 1835 prediction, the comet reached its closest approach to the sun too late to agree with theory.

This several days of error could not be satisfactorily explained by the existence of the hypothetical "resisting medium" or "luminous ether" since the comet was delayed and not speeded up on this return. The laws of motion and gravitation had by now been verified beyond a doubt. Yet Halley's comet appeared to violate the very laws upon which its periodic nature was first surmised. However, another possibility existed. In 1861, James C. Watson in his book, *Treatise on Comets,* had written, "It is possible that the orbit of Neptune does not mark the confines of the planetary system, but that there may be other planets still remote, which, in due course of time, may change the motions of the comet, and which, in case they are not sooner revealed, may finally cause it to fail to appear in accordance with prediction."

In 1846, the discovery of Neptune was itself a consequence of Newton's law of universal gravitation and a spectacular demonstration of its validity. Orbital irregularities had been noted in the planet Uranus, previously discovered in 1781. The possibility of an undiscovered planet that might be perturbing the orbit of Uranus was investigated independently by John Couch Adams (1819-1892) in England and by Urbain Jean Joseph Leverrier (1811-1877) in France. As a result of their calculations, the position of this theoretical planet was determined, and on the first night of searching it was found by the German astronomer Johann Galle (1812-1910), exactly where predicted.

Some slight irregularities in the orbit of Neptune prompted a search for an even more distant planet. In 1930, Clyde Tombaugh of Lowell Observatory in Flagstaff, Arizona, discovered Pluto, the eccentric wanderer of the skies whose orbit, at the time of Halley's 1986 perihelion, will place it inside that of Neptune. It is now theorized that Pluto may not be a true planet at all, but rather a wayward satellite of Neptune. Also, a body as small as Pluto would not noticeably affect a giant planet like Neptune. In an article entitled "The Great Unexplained Residual in the Orbit of Neptune" published in September 1970 in the *Astronomical Journal,* Dennis Rawlins suggested that the error in the calculated orbit of Neptune may be "...due to an alien perturbation." Therefore, the possibility that an undiscovered planet is perturbing Neptune's motion and perhaps the orbit of Halley's comet as well would tend to warrant serious consideration.

In April 1972, Joseph L. Brady of the Lawrence Livermore Laboratory in California wrote a paper entitled "The Effect of a Trans-Plutonian Planet on Halley's Comet." He found that he could reduce the error in the predicted returns of the comet by assuming the presence of a hypothetical planet of a mass three times that of Saturn with a period of 464 years and moving in a retrograde orbit which is inclined by 60 degrees to the plane of the rest of the solar system. He predicted that such a massive object would be quite large and bright, visible telescopically, and located in the area of the constellation of Cassiopeia. Several independent searches for this hypothetical planet, popularized as

Planet X, were made. But no new tenth planet was found, despite careful observations at the predicted position.

It was suggested that such a large mass would have a profound effect upon the orbits of the outer planets, shifting them out of the plane of the ecliptic. Instead, a modern theory of comets was postulated to suggest a nongravitational force as a cause for the comet's irregular motion. Also, a review of past apparitions by Tao Kiang at Dunsink Observatory in Ireland suggested different dates from those adopted by Dr. Brady for his calculations.

Still, the anomalies of Neptune's orbit have not been satisfactorily explained and the current theory of comets is as yet unproven. The possibility of an undiscovered planet far beyond the known boundaries of our solar system, altering the orbits of comets, cannot be dismissed.

CHAPTER 6

Back in Time

> **In the seventh year of Chhin Shih-Huang-Ti [240 B.C.] a comet first appeared in the east....**
>
> **—the *Shih Chi***

Once the periodicity of Halley's comet was well established, the question naturally arose which of the numerous comets observed throughout history represented this returning apparition. It was Halley himself who first carried the identification of his comet back to the year 1456. The returns in 1531, 1607, and 1682 gave him intervals of 76 and 75 years. Following his initial prediction, Halley examined catalogues of past apparitions and noticed that three others had shown at roughly the same interval of time: in the year 1305 at about Easter, during the year 1380, and in 1456 during the month of June. Absence of accurate observations, however, prevented him from calculating an orbit and his identifications prior to 1456 were later shown to be wrong.

Alexandre Pingré (1711-1796) confirmed Halley's suspicion that the comet had previously returned in the year 1456. Pingré was a French astronomer noted for his monumental two volume opus, *Cométographie*. Published in 1783-84, this work was long a source book for cometary information, and contained observations of comets recorded in both European and Chinese chronicles. During his research for this book, Pingré found definite observations of the position of the comet of 1456, one observation obtained at Vienna, the other at Rome. It was a very bright comet with a tail 60 degrees long. Pingré calculated an orbit based upon these independent observations and found that it corresponded to that previously calculated for Halley's comet.

Johann Burckhardt (1773-1825), a mathematician and astronomer who was born in Leipzig and eventually became a naturalized French citizen, wrote many articles on comets. His was the next positive identification. Investigating the orbit of a comet observed in the year 989 in China and also mentioned in several Anglo-Saxon chronicles, he found that the elements of the orbit calculated from these observations bore considerable resemblance to those of Halley's comet.

Accurate observational data is needed in order to calculate a comet's orbit. It was obvious that there was little of astronomical value in the early European cometary records. Superstitious dread of comets meant that their possible malign significance was more important to past observers than their exact position in the sky.

The most trustworthy data before the fifteenth century regarding cometary appearances were to be found in the records of the Chinese observers. There had been a long history of accurate astronomical observations in China, and records were kept in a much more systematic manner than in the chronicles of Europe.

It is believed that during his reign (about 2600 B.C.) the emperor Hoang-Ti had an observatory built mainly for the purpose of correcting the calendar. A mathematical tribunal was instituted to promote the astronomical sciences, especially the prediction of eclipses, and the members of this tribunal would forfeit their lives if their predictions were inaccurate. During the reign of Chung K'ang (about 2000 B.C.) two astronomers, Hsi and Ho, were put to death because of an eclipse which they did not foresee.

Numerous observations of comets extending from between one and two thousand years before the Christian era were recorded by the Chinese. These observations were unavailable to the European astronomers, however, until the arrival of the Jesuit missionaries in China in the early years of the seventeenth century. Under the leadership of Father Matteo Ricci (1552-1610), the missionaries brought the Chinese court the word of European astronomy along with the Word of God. The Jesuits supplied the Chinese with the foundations of modern calendrical science so important for the life of the nation. But the information flow worked both ways. Through the missionaries' communications with the West, China's wisdom became known to Europe, including extensive astronomical observations reaching back over two millennia.

Father Antoine Gaubil (1689-1759), a French Jesuit mathematician and astronomer, became a missionary to China in 1723. He wrote a history of Chinese astronomy and provided compilations of many of their observations in works published in Paris in 1732. Other individuals gradually added to the store of Oriental observations available in the West. These included Achille Dionis du Séjour (1734-1794), a French astronomer who also sought to demonstrate the improbability of a comet's ever colliding with the earth.

Later, Édouard Biot (1803-1850), a historian, noted in his communication to the *Connaissance des Temps* for 1846 that the appearance of 224 comets had been recorded by the Chinese. He translated the majority of these observations from the *Wên Hsien Thung Khao (Comprehensive Study of Documents and Records)*, completed by Ma Tuan-Lin in the year 1254, and from the supplement to the same work, the *Hsü Wên Hsien Thung Khao*, written by Wang Chhi in 1586 and later revised and rewritten during the middle of the eighteenth century.

The work by Ma Tuan-Lin was described by the Oriental philologist Abel Rémusat (1788-1832) as "the finest monument of Chinese literature, a vast collection of memoirs on all sorts of subjects, a treasure of erudition and criticism, in which all the materials that Chinese antiquity has left us...will be found, brought together, classed and discussed, in admirable order, method, and lucidity." This astounding compendium is composed of 100 volumes, divided into 348 sections or chapters. The astronomical portion consists of 17 chapters, numbers 278 to 295 inclusive, and among its subjects are the Chinese division of the visible heavens into thirty-one parts, accounts of the sun, moon, and the five visible planets, eclipses of the sun, the appearance of sun spots, eclipses of the moon, observations of comets, records of shooting stars or meteors, and accounts of extraordinary halos and rainbows.

Working from these Chinese sources, Paul Auguste Laugier (1812-1872), an astronomer at the Paris Observatory, identified three additional appearances of Halley's comet. In 1843, in a report submitted to the *Comptes Rendus*, Laugier determined the parabolic elements of the comet of 1378 on the basis of a translation by Biot which recorded that comet's progress through the heavens. Later, in 1846, relying upon a translation of Ma Tuan-Lin's compilation of cometary observations, Laugier identified the returns in 760 and 451 as other previous apparitions of the comet.

The first person to make extensive use of these Chinese comet observations, however, was the Englishman, John Russell Hind (1823-1895), who used them to trace the returns of Halley's comet to before the Christian era (Fig. 6-1). Hind was born at Nottingham, England. He showed an interest in astronomy at an early age and contributed a number of astronomical notes to his local newspaper when he was only sixteen years old. In 1840, he was appointed to a post at the Royal Observatory in Greenwich in the newly formed Magnetical and Meteorological Department, where he stayed

Fig. 6-1. John Russell Hind (1823-1895). His researches into the past records of comets enabled him to trace Halley's appearances back to 12 B.C. He was the first to associate Halley's comet with William the Conqueror and the Battle of Hastings in 1066.

until 1844. He then became employed as an observer at a private observatory in Regent's Park, London, owned by a successful businessman, Mr. George Bishop.

In 1847, Hind discovered two asteroids which he named Iris and Flora. Over a nine-year span, he discovered ten asteroids and three comets, besides *variable stars*, that is, stars that vary in brightness, and a variable nebula in Taurus. His main work, however, was in connection with comets. He was a prolific contributor to scientific journals and wrote a popular book on the subject. In 1851, Hind was made a fellow of the Royal Astronomical Society and eventually became its president. His computational skill and diligence as an observer gained him a wide reputation and a tangible reward in the form of an annual pension of two hundred pounds "in consideration of his contribution to astronomical science by important discoveries."

Of the personal details of his life, we know that he was an excessively nervous individual and of a very retiring disposition. Married in 1846, he fathered six children. Following his death from the complications of heart disease, Hind was remembered as one who "worked more for science' sake than for the approbation of his fellow-men or for his own pecuniary advancement."

In tracing backward the various apparitions of Halley's comet through the centuries, Hind relied upon the *Cométographie* of Pingré, the works of Hevelius, Stanislas Lubieniecky's *Theatrum Cometicum*, which claimed to list all the comets seen or recorded from the deluge to 1665, the contributions of Édouard Biot, and Ma Tuan-Lin's catalogue of comets. Hind's method involved assuming approximate intervals between appearances of the comet and searching the past records of comets for those with elements similar to those of Halley's comet.

The apparition of 1378 had been previously confirmed by Laugier in 1843, so Hind turned his attention to an assumed return in 1301. In that year a great comet had appeared with a bright and extensive tail stretching from the west towards the eastern parts of the heavens. It was seen in Europe as far north as Iceland. In China the comet was first seen on 16 September and remained visible until 31 October. Laugier, using a combination of Chinese observations and a description by Friar Giles in Europe, had obtained orbital elements which differed from those expected for Halley's comet. Hind, however, completely rejected the European narrative and relied solely upon the Chinese account, which he felt agreed extremely well with the expected elements. It is now known that he was correct in this assumption.

As he continued his researches, Hind became convinced that "the preceding return of the comet took place, I think, in 1223, in the month of July, shortly before the death of Philip Augustus [of France]." In this identification, however, Hind was mistaken. His error was the result of his finding no mention of this comet by the Chinese, and the European records, including the account by Pingré, were vague. Also, this revolution was one of the longest on record, approximately seventy-nine years, and such a wide variance of the period of Halley's comet doubtlessly put Hind off the track.

Hind was more confident about the next return of 1145, which was witnessed by both European and Oriental historians. The Chinese chronicles describe the comet as being first seen in the eastern sky on 26 April, which was several days after the perihelion passage of 19 April established by Hind. It was noted to be bluish-white in color and was lost to view in June.

In the year 1066, "a very grand and remarkable comet," according to Hind, astonished Europe. During its sixty-seven day visibility, it was minutely described in the Chinese annals. Hind found that the elements of this comet, derived from observation, did not coincide with those which he had calculated for Halley's comet. But he questioned whether "in the lapse of so many centuries may not the planetary perturbations have produced alterations in the elements...?" Hind was correct and, although his derived elements later had to be modified, he was the first to identify this apparition immortalized in the Bayeux Tapestry, celebrating William the Conqueror's victory at the Battle of Hastings, with Halley's comet.

The orbit of the comet of 989 had been previously computed by Johann Burckhardt, who noted its similarity to that of Halley's comet. Hind determined that perihelion passage would have occurred about 12 September.

The return of the comet in 912 was erroneously conjectured by Hind as having a perihelion passage toward the beginning of April. He based this determination upon precise Chinese records. These observations conflict, however, with those of the Japanese which noted a comet's appearance in July. In 1972, Kiang, in his article "The Past Orbit of Halley's Comet," suggested that what the Chinese saw was a fragment of the comet which preceded the main body by three months.

One of the most spectacular returns ever occurred in 837. It was observed during March and April by the Chinese astronomers who recorded a tail length exceeding ninety degrees. The comet approached to within 4 million miles of the earth in early April. Pingré had previously calculated the elements of this body, which showed a general similarity to those of Halley's comet. However, Pingré as well as Hind incorrectly assumed passage on 10 April through the *ascending node*, that is, the intersection of the comet's orbit with the ecliptic in a northward direction, when in actuality the comet was traveling southward through the *descending node* at about this time. Hind stated that "the comet of 837 which figures in our catalogues was, therefore, in all probability, different from Halley's, though the position and distance in perihelion and direction of motion were the same, and the inclination of the orbit...not very widely different for the two bodies." Hind incorrectly assumed the appearance of a "guest-star" or nova recorded for 29 April as the actual return of the comet, with a perihelion occurring earlier in the month. The Chinese noted a total of five objects appearing in the heavens in 837. In all likelihood the first object seen was Halley's comet.

We have previously discussed how Laugier confirmed the return in 760 using the unusually precise observations made in China, where the comet was visible for more than fifty days. In Europe during the twentieth year of the emperor Constantine V, a comet like a brilliant beam of light was visible for about thirty days, appearing first in the east and subsequently in the west. According to Hind, "the appearance of our comet in 760 is a matter little short of certainty."

Three objects were sighted in the year 684 and recorded in both Chinese and Japanese chronicles. According to the account compiled by Ma Tuan-Lin, a comet was seen in the western skies during the months of September and October. This brief report does not supply enough information necessary for an orbit calculation, however. Later, in 1905, Edward Knobel (1841-1930), a linguist and former president of the British Astronomical Association, commented on the astronomical observations recorded in the *Nihongi*, a history of ancient Japan. He notes that in the "thirteenth year of Temmu, A.D. 684, autumn, seventh month, 23rd day, a comet appeared in the northwest more than ten feet long." This description corresponds to a date of 7 September and the object noted was probably the returning apparition rather than two others sighted later in the year.

Hind erred in his next determination by over a year and a half. He stated in 1850 that "the previous return should have taken place about the year 607, and the Chinese annals have several comets in that year. I find, by actual computation, that none of them present any decided indications of identity with the one which forms the subject of these remarks, and I am therefore inclined to fix its reappearance in the following year, 608, when a comet is mentioned by Ma Tuan-Lin." The mistake was made because of conflicting Chinese observations and Hind's reliance upon "close agreement of intervals."

The comet which appeared in 530 was believed by both Newton and Halley to be a previous return of the "Great Comet" of 1680 which they had both witnessed. Hind was convinced that these comets were not identical and suggested that perhaps the apparition of 530 was a reappearance of Halley's comet. "Yet this inference is necessarily open to considerable doubt, and I am very far from insisting upon it," he said. Hind had the correct comet but was slightly in error regarding the date of perihelion, for he was misled by an incorrect month given in the secondary Chinese source he was using. According to the primary source, the *Wei Shu*, the comet was observed during the months of August and September and disappeared on the 27th day of the latter month.

It was an interval of seventy-nine years and three months to the next apparition in 451, the longest on record. Laugier had previously shown this comet to be that of Halley, finding a significant agreement between the Chinese observations and its calculated position in the heavens. Both in Europe and in China the comet was initially discovered on 10 June and followed in its course from the Pleiades into the constellations of Leo and Virgo.

According to Hind, "In October 373, twenty-fourth day, the Chinese mention a comet in Ophiuchus and Serpens...The account is too vague, however, to allow of any definite conclusion."

Hind deducted a period of 78 years and found recorded in chapter 286 of Ma Tuan-Lin's catalogue the path of a comet "exactly represented by the orbit of Halley's Comet." Hind compared its track among the stars with the orbits of the comets in 451, 760, and 1456 and found them sufficiently similar to justify confirmation of its reappearance in the year 295.

A certain identification of the comet was made for the year 218. Dion Cassius (ca. 155-ca. 230), a Roman politician and historian, described it as a very fearful star, extending its tail eastwards from the west. The Chinese noted its intense brilliance and described its initial appearance as a morning star in the east with its gradual passage through the constellations of Auriga, Gemini, and Leo. Hind fixed the date of perihelion passage for 6 April.

Few comets were recorded about the expected time of the next apparition. But according to the Chinese records, a comet of a bluish-white color appeared in the eastern sky on the morning of 27 March in the year 141. A precise observation made on 16 April enabled the elements of an orbit to be computed which were "not very widely different from those of Halley's" according to Hind. Traversing the constellations of Auriga, Gemini, Ursa Major, and Leo, the comet finally disappeared from view.

According to Hind, "The preceding apparition took place either in the summer of the year 65, or early in 66. Two comets are found in the Chinese annals." He decided in favor of the latter apparition, which proved to be the correct choice. The *Hou Han Shu (History of the Later Han Dynasty)* recorded that "in the twelfth month, on the day Ou Tse, a new star appeared in the east sky." This date corresponds to 31 January, which was close to Hind's calculated perihelion passage.

In the year 12 B.C., or -11 according to astronomical notation, a comet seemed to be suspended over the city of Rome. The Chinese first noted its appearance on 26 August, in Gemini, and carefully observed it for 56 days, when it disappeared into Scorpio. Hind concluded that the elements of this orbit left no doubt as to its identity.

At this point Hind concluded his search since "the accounts of comets become so vague that it would be vain to attempt to carry the inquiry into more remote antiquity." He also expressed his indebtedness to "the records preserved in the annals of the various reigning dynasties in China."

During his researches, Hind discovered a discrepancy in the path of a comet contained in Biot's catalogue of comets observed

Fig. 6-2. John Williams (1797-1874). In 1871, he compiled and published at his own expense a catalogue of 373 observations of comets which later provided confirmation for mathematical calculations of past apparitions of Halley's comet.

in China. He communicated this fact to John Williams (1797-1874), the assistant secretary of the Royal Astronomical Society (Fig. 6-2). Williams undertook a systematic examination of Biot's catalogue and quickly found that, although very accurate in its details of cometary appearances, it was more incomplete than was expected. About 150 observations of comets recorded in the opus by Ma Tuan-Lin and in the *Shih Chi (Historical Records)* by Ssuma Chhien and his father Ssuma Than were not noted by Biot. Williams felt that a catalogue comprising all the cometary observations contained in these two Chinese works, translated from the original and arranged chronologically, might be of service to astronomers. Consequently, he obtained a total of 373 observations, and in 1871 he had published at his own expense a quarto volume of about 250 pages entitled *Observations of Comets from B.C. 611 to A.D. 1640, extracted from the Chinese Annals*. This comet catalogue provided observational support for calculations performed by a pair of Englishmen in the early years of the twentieth century and made it possible for them to push the date of confirmed appearances of Halley's comet to an even earlier time.

Philip Cowell (1870-1949) was born the second son of a barrister in Calcutta, India (Fig. 6-3). In 1896 he was appointed chief assistant at the Royal Observatory at Greenwich and was elected a fellow of the Royal Society in 1906. His name is perpetuated in "Cowell's method," which is a simplified way to calculate the perturbations of celestial bodies.

Andrew Crommelin (1865-1939) was born in Northern Ireland and educated at Cambridge (Fig. 6-4). Like Cowell he became an assistant at the Royal Observatory. He later was to serve as president of the Royal Astronomical Society. In 1919 his photographs of the solar eclipse during an expedition to Brazil helped confirm Einstein's revision of the Newtonian universe by establishing that light is deflected by gravitation. Crommelin's principal contribution to astronomy, however,

Fig. 6-3. Philip Cowell (1870-1949). He derived a simplified method to calculate the perturbations upon the orbit of Halley's comet and with A.C.D. Crommelin used it to brilliantly predict the perihelion passage of the comet in 1910.

Fig. 6-4. Andrew C.D. Crommelin (1865-1939). He served as president of the Royal Astronomical Society from 1929-1931. Together with Philip Cowell, he authored a series of papers which extended Hind's list of previous apparitions of Halley's comet back to 240 B.C.

involved comets and minor planets. His orbit calculation of a comet seen in 1928 eventually resulted in that comet being named after him. Both he and Edmond Halley are among the few individuals in which a comet has been named after the computer of the orbit rather than after its discoverer.

In a series of papers contributed in 1907 and 1908 to the *Monthly Notices of the Royal Astronomical Society*, Cowell and Crommelin succeeded in tracing the return of Halley's comet back to 240 B.C., both extending Hind's list and correcting several errors in his predicted perihelion passages. To assist them in their calculation—which, despite Cowell's simplified method, were still arduous—the two British men had available the services of three volunteer computers, Dr. Smart, Mr. F.R. Cripps, and Mr. Thomas Wright.

The goal of their efforts was to carry back "the calculation of the perturbations as far as possible, and seeing whether a sufficiently accurate correspondence existed between the conjectured and calculated dates." For determining the observed dates, they relied upon the descriptions by Pingré in his *Cométographie* and the translations from the Chinese by Williams. For the calculated time of perihelion, they determined the perturbations caused by Venus, Earth, Jupiter, Saturn, Uranus, and Neptune upon the comet during each return. They confirmed each return before proceeding backwards to the next. Because of the enormity of the calculations, some assumptions had to be made, such as a constant eccentricity of the comet and a uniform change in the perihelion and node from revolution to revolution. Still, their research was a model of its kind.

Philippe Gustave de Pontécoulant had previously been awarded a prize in 1829 by the Academy of Sciences of Paris for closely determining the date of the 1835 perihelion passage of Halley's comet. Later in 1864, writing in the *Comptes Rendus* of the French Academy, he determined the effect of planetary perturbations upon the comet by calculating the dates of perihelion for its five previous appearances back to the year

1531. Cowell and Crommelin began their calculations at this point. The apparitions of 1456 and 1378 had previously been confirmed by Pingré and Laugier, respectively; therefore, the first corroboration of a return calculated by Hind was the apparition of 1301. "The discordance is so small that we are justified in accepting Hind's result with absolute confidence," wrote Cowell and Crommelin.

However, they found that Hind's identification of the comet of July 1223 with Halley's was erroneous, his determination being ten months too late. The calculations of Cowell and Crommelin agreed with a description of a grand comet that appeared in September of the year 1222 during the reign of Ning Tsung as translated by Williams.

"Hind's identification of the comet of 1145 with Halley's is correct," declared Cowell and Crommelin. With regard to the previous return in 1066, they found that the "identification of the famous comet that preceded the Norman Conquest with Halley's comet is fully established."

Burckhardt had investigated the return in 989, but Hind had determined the time of perihelion. Cowell and Crommelin confirmed both the comet as being Halley's and Hind's date of the comet's closest approach to the sun.

Their calculated perihelion for the return in 912 differed from that obtained by Hind, who based his calculation solely on the Chinese observations, being unaware of the Japanese records of this event. Cowell and Crommelin found that their date was "nearly 4 months later than Hind's date, a larger quantity than is likely to arise from error in our calculation."

The return in 837 was determined by the two British astronomers to have been the first object observed in that year and not the second as thought by Hind. Cowell and Crommelin confirmed the return in 760. Their computations indicated that Hind's identification was correct for the return in 684, although their calculated perihelion date was over a month later than his, "which is a reasonable discordance for that remote epoch."

Hind was shown to have chosen the wrong comet in 608, for Halley's comet appeared one and one-half years earlier according to the calculations of Cowell and Crommelin, but "the observations of 607 are in a decided tangle." In 1972 Kiang noted, "The return of Halley's Comet in 607 must rest on rather slender observational support pending further clarification of the records." The next two returns in 530 and 451, identified by Hind and Laugier respectively, were both confirmed.

Cowell and Crommelin calculated the next perihelion passage to have occurred on 7 November 373, which was not significantly different from Hind's determination, although it lacked observational confirmation. It was later noted in 1911 by Kiyotsugu Hirayama (1874-1943) of the Astronomical Observatory in Tokyo that discrepancies existed in the compilations of Ma Tuan-Lin. Hirayama, whose hobby was oriental classics, researched the original records of the Chin and Sung dynasties and found that the comet which appeared in the second year of Ning-Khang, in the second month, was most likely Halley's comet. Using the method of Cowell and Crommelin to calculate the perturbations of Jupiter and Saturn, he derived a perihelion date of 13 February 374.

The remaining apparitions determined by Hind were confirmed back to 12 B.C. by Cowell and Crommelin, but they were also able to extend Hind's list back even earlier. From a description in Williams" catalogue, they noted that a comet appeared in August and September of 87 B.C. They were fairly certain that this was a previous return. The next appearance would most likely be in 163 or 164 B.C., but no definite Chinese observation could be found, although Pingré mentioned that in 163 B.C. "the Sun was seen at night." The two astronomers were confident of the next return in 240 B.C. Although the date of the appearance as recorded in the *Shih Chi* and in the secondary source by Ma Tuan-Lin did not agree with their calculated date of January 239 B.C., they felt that the discrepancy was not significantly out of line.

For the next two passages Cowell and Crommelin could find no comets that might be Halley's, although, "in the second year of the Emperor Ching Ting Wang (B.C. 467)," according to Williams, "a comet was seen." There is no mention of the time of year nor of the path of the comet through the heavens. Nevertheless, a return was expected about that time, and in Greece, where it was also seen, Aristotle recorded that a great meteorite fell while the comet was visible in the sky. Since meteor showers are associated with Halley's comet, this fact strengthens the conjecture that this was another visit by that famous comet.

Concluding their research, Cowell and Crommelin stated that "we have carried the comet with fair certainty back to B.C. 87, with some probability back to B.C. 240; at this point we are brought to a standstill by the complete absence of earlier observational material."

An update of Chinese observations was made in 1962 by Ho Peng Yoke of the University of Malaya, Singapore, and later of Cambridge in England. Writing in the publication *Vistas in Astronomy*, Ho noted that Williams" catalogue contained many errors, which included misinterpretation of the original textual material, neglect of changes in the Chinese calendar, and mistakes in applying his system for converting Chinese dates to Julian dates. He also brought to light incompleteness and inaccuracies in the compilation by Ma Tuan-Lin as well as in other secondary sources of cometary observations.

Ho examined the primary Chinese sources, supplemented them with Korean and Japanese observations, and compiled 581 different listings of comets and novae from the fourteenth century before the Christian era up to 1600. This information was of value in one further refinement in the history of the past orbit and perihelion passages of Halley's comet.

Tao Kiang of the Dunsink Observatory in Castleknock, Republic of Ireland, calculated the perturbations of the orbit of Halley's comet during its previous twenty-eight revolutions. His results were published in the *Memoirs of the Royal Astronomical Society* for 1972. Unlike Cowell and Crommelin's, Kiang's computer was of the electronic kind (an IBM 1620), and he had available to him Ho's detailed oriental researches.

Kiang determined the time of perihelion passage, if at all possible, from observations rather than from calculations. The remaining five elements which determine the comet's orbit were allowed to vary under the perturbative influence of all the planets. Whereas in Cowell and Crommelin's study the perturbations in the longitudes of the ascending node and the perihelion point were merely assumed rather than calculated, Kiang computed the perturbations in the elements "with full rigour." These and other refinements were made, and the times of perihelion passage determined.

Finally, in 1981, Kiang together with Donald K. Yeomans of the Jet Propulsion Laboratory in Pasadena, California, re-examined the Chinese material in the light of additional information. The two men included nongravitational effects upon the comet and made refinements to the previous perihelion passage dates computed by Kiang. Their results, together with those obtained earlier by Hind and Cowell and Crommelin, are shown in Table 6-1. It is not expected that a more accurate determination of the dates of past returns of Halley's comet can be obtained.

RETURNS TO PERIHELION OF HALLEY'S COMET

Return	Hind (1850)	Cowell and Crommelin (1910)	Yeomans and Kiang (1981)
-1		1910 Apr 16	1910 Apr 20
-2	1835 Nov 14	1835 Nov 15	1835 Nov 16
-3	1759 Mar 10	1759 Mar 12	1759 Mar 13
-4	1682 Sep 12	1682 Sep 14	1682 Sep 15
* -5	1607 Oct 26	1607 Oct 26	1607 Oct 27
-6	1531 Aug 25	1531 Aug 25	1531 Aug 26
-7	1456 Jun 8	1456 Jun 2	1456 Jun 9
-8	1378 Nov 8	1378 Nov 8	1378 Nov 10
-9	1301 Oct 22	1301 Oct 23	1301 Oct 25
-10	1223 Jul 9	1222 Sep 10	1222 Sep 28
-11	1145 Apr 19	1145 Apr 19	1145 Apr 18
-12	1066 Apr 1	1066 Mar 25	1066 Mar 20
-13	989 Sep 12	989 Sep 15	989 Sep 5
-14	912 Apr 1	912 Jul 19	912 Jul 18
-15	837 Apr 6	837 Feb 25	837 Feb 28
-16	760 Jun 11	760 Jun 10	760 May 20
-17	684 Oct 18	684 Nov 26	684 Oct 2
-18	608 Oct 19	607 Mar 26	607 Mar 15
-19	530 Nov 3	530 Nov 15	530 Sep 27
-20	451 Jul 3	451 Jul 3	451 Jun 28
-21	373 Nov 3	373 Nov 7	374 Feb 16
-22	295 Apr 1	295 Apr 7	295 Apr 20
-23	218 Apr 6	218 Apr 6	218 May 17
-24	141 Mar 29	141 Mar 25	141 Mar 22
-25	66 Jan 26	66 Jan 26	66 Jan 25
-26	12 B.C. Oct 18	12 B.C. Oct 8	12 B.C. Oct 10
-27		87 B.C. Aug 15	87 B.C. Aug 6
-28		163 B.C. May 20	164 B.C. Nov 12
-29		240 B.C. May 15	240 B.C. May 25

Table 6-1

* Gregorian calendar dates are given for returns -1 to -5; Julian calendar dates for all other returns.

CHAPTER 7

The Historical Comet

Old men and comets have been reverenced for the same reason: their long beards, and pretences to foretell events.

—Jonathan Swift, *Thoughts on Various Subjects*

The comet with the eponym of Edmond Halley is unique in that it has a history. Like other comets of the past, it was an object of superstition and often feared as a harbinger of pestilence, famine, war, political turmoil, and other adverse news. Indeed, the word *disaster* literally means "bad star." Such apparitions were often believed not simply to foretell events but also to possess the power of influencing them directly. Unlike others of its kind, this celebrated comet is a bright celestial object that is easily spotted without a telescope, that returns after a comparatively short interval, and that has been repeatedly witnessed for at least twenty centuries. Its many returns necessarily coincided with the natural and historical events befalling earth in those years of its passage. But it was only after the periodicity of Halley's comet had been established and its many former returns confirmed, that it began to be associated in the popular mind with significant and pivotal events in Western history.

To be sure, during many of the comet's appearances nothing of earthly import was occurring. And connections can be established between significant historical events and the visitations of other comets just as easily. But no records of important events associated with a comet seem as remarkable or as extensive as those coinciding with the appearance of Dr. Halley's famous comet.

What follows is a catalogue of a few of the many historical events which this comet has witnessed—and for which in many cases it has been held responsible.

The earliest confirmed date of an appearance of Halley's comet is 240 B.C. Unfortunately, not a great deal was happening of world historical importance, although in the previous year Rome did take a giant step toward empire by crushing the Carthaginians in a naval battle near the Aegadian Islands off western Sicily. The republic exacted from Carthage an indemnity which included the cession of Sicily and the islands around it. With the end of the First Punic War, Rome was on its way to world domination. Halley's comet watched.

In the years near the return of Halley's comet in 12 B.C., several significant astronomical events occurred. Among these celestial happenings is the one that announced the beginning of the Christian era—the biblical star of Bethlehem. The account by Matthew the Evangelist reads, "And lo, the star which they had seen in the East went before them, till it came to rest over the place where the child was. When they saw the star, they rejoiced exceedingly with great joy...." Using this brief and undetailed description, astronomers and biblical scholars have long sought to identify the star that guided the Magi.

The first problem is in the word *star*. Matthew uses the Greek word *aster*, which at the time was applied to almost anything in the nighttime sky. There were fixed or true stars which remained apparently motionless in the heavens, wandering stars or planets, new stars or novas, falling stars or meteors, and hairy stars or comets. Whatever the natural basis of this particular celestial spectacle, Matthew records that it was so extraordinary that it caused the Magi, whose home was either Chaldea or Persia, to set out on the long journey to Bethlehem, and that this heavenly display continued long enough to allow them time to follow it.

The word *magi* is derived from the Greek *magoi*, which in turn is derived from a Persian word meaning astrologer. The Magi were those select few whose duty it was to predict and interpret changes among the stars. Since the heavens supposedly revealed the destiny of nations, it was the Magi's job to record any celestial signs that might announce the birth or death of heroes and conquerors.

The exact year of the birth of Jesus of Nazareth remains uncertain. Estimates range from 12 B.C. to 1 B.C. Lack of a precise date means that all attempts to provide an exact astronomical explanation for the appearance of the star are ultimately speculative. The theories advanced include: an unusual juxtaposition or conjunction of the plants, the bright planet Venus, the sudden appearance of a nova, a bright meteor, a comet, and some unique supernatural event. And there are those who explain the star as nothing more than a literary device that Matthew uses to emphasize the birth of a great man. He fabricated the star, they say, because a celestial sign was appropriate for this significant event in human history.

The most widely accepted explanation for the star was that advocated by Johannes Kepler in the first decade of the seventeenth century, who suggested that it was the result of a rare triple conjunction of Jupiter and Saturn in the constellation Pisces in 7 B.C. During this year, three separate close passages of the two planets occurred. Babylonian astrologers of the period saw Jupiter as the royal planet, while Saturn governed the fate of the Jews, and Pisces was traditionally identified with the destinies of Israel. Such a conjunction in that sign would necessarily signify the rising of some mighty master of the Jewish race.

An alternative tenable explanation is that the wise men were guided by a comet. Comets, too, were regarded as divine messengers—usually portents of calamity, doom, and the death of kings, but also sometimes as positive omens of victory, bounty, and the birth of kings.

The description that the star "came to rest over the place where the child was" would agree with a cometary explanation. The historian Dion Cassius recorded the appearance of a comet in 12 B.C., just before the death of Marcus Agrippa, a Roman general, during the consulate of M. Valerius Messala and P. Sulpicius Quirinus. This was Halley's comet. As noted by Dion Cassius, the comet seemed to be suspended directly over the city of Rome. A plot of its course, obtained from Chinese records, indicates that the comet crossed the zenith of Bethlehem, appearing to hover over that city. Since the latitude of Rome differs only 10 degrees from that of Bethlehem, the position of the comet in the sky would have been similar for the people of both cities.

Giotto di Bondone (1267-1337), a Florentine pioneer of naturalistic painting, covered the interior walls of the Scrovegni Chapel in Padua with frescoes illustrating the life of Christ. One of the scenes, *The Adoration of the Magi*, which he completed in either 1303 or 1304, depicts the star of Bethlehem as a comet (Fig. 7-1). Giotto almost certainly was using as his model Halley's comet, which had appeared in 1301. He was unaware, of course, that over thirteen centuries earlier this same comet may in fact have witnessed the birth of the Christ child.

The Gospel of Matthew was written, the scholars tell us, after the fall of Jerusalem in A.D. 77. If the star of Bethlehem was more a literary device than an astronomical fact, Matthew may well have been influenced by the appearance of Halley's comet, which he doubtlessly witnessed in the year 66. It is interesting that a comet which appeared just prior to the birth of Mohammed was subsequently believed by his followers to have

Fig. 7-1. *The Adoration of the Magi* by Giotto di Bondone (1267-1337). This frescoe decorates the interior of the Scrovegni Chapel in Padua and depicts the star of Bethlehem as a comet. The apparition of Halley's comet in 1301 served as Giotto's model. The European Space Agency is sending its *Giotto* spacecraft to investigate Halley's comet during 1986.

announced his birth, in the same way as had the star that appeared at the birth of Christ.

The return in 66 was also probably the same comet that terrified the Jews as it hung over the city of Jerusalem. Flavius Josephus, a Jewish historian and soldier of the first century, wrote in his *Bella Judaeorum (The Jewish War)*, "Amongst other warnings, a comet, of the kind called *Xiphiae*, because their tails appear to represent the blade of a sword, was seen above the city for the space of a whole year." Josephus rebuked his countrymen for listening to false prophets while so notable a sign was present above their heads. Shortly afterwards, the city was put under siege and eventually destroyed by the Romans under Titus Vespasian. Actually, there were two comets that were seen during this time, one during the summer of 65 and Halley's, which appeared early in 66. Their consecutive appearances probably accounted for the reference made to the year's length of time that the comet was seen.

For Dion Cassius, Halley's comet during its return of 218 looked like a "fearful flaming star" announcing the death of Emperor Macrinus. But the return of 451 was to prove even more fateful for the history of Rome.

In 450 Emperor Marcian had refused to pay tribute to the Huns. Their infamous leader, Attila, known as "the Scourge of God," thereupon invaded Italy. The comet of Halley appeared in the skies over Europe during the late spring of 451, and shortly afterwards, near the town of Châlons in the northeast of France, the Roman armies under General Flavius Aëtius defeated Attila. Numerous historians and chroniclers of the day agreed that the victory had been announced by the comet. Though it was regarded as a battle between the forces of Christianity and heathendom, this victory is most memorable as the last important one ever gained by Imperial Rome. Attila died in 453, and after his death the Hunnish Empire collapsed.

In the Dark Ages that followed the fall of Rome, the superstitious dread with which Halley's comet was regarded swayed many a ruler. It was instrumental in forming the policies of Louis le Débonnaire in 837, and some thought that it was a warning of his death. Louis (778-840), the youngest son of Charlemagne, had inherited the Frankish throne in 814. An anonymous chronicler of his reign described the comet and its effect upon him: "During the holy days of Easter, a phenomenon which is always ominous and a carrier of bad news appeared in the sky. As soon as the Emperor, who always paid great attention to such events, had noticed it, he allowed himself no rest....'A change of reign and the death of a prince are announced by this sign,' he told me. He consulted the bishops, who advised him to pray, build churches and found monasteries. Which he did. But he died three years later."

Bayeux, France, lies near the Normandy coast, not far from the site of the Allied invasion in June of 1944. Preserved within a museum in this town is a world famous embroidery of colored wool on coarse linen 231 feet long by 19 1/2 inches wide. This is the priceless Bayeux Tapestry or *La Tapisserie de la Reine Mathilde*, which depicts in 58 dramatic scenes the Norman conquest of England in 1066. Though neither strictly a tapestry nor the work of Queen Matilda, it is recognized as both a unique historical document of the age of chivalry and a valuable contribution to medieval art.

One of the scenes shows several Englishmen huddled together, pointing their fingers and gazing in terror and amazement at the comet in the sky. The legend over the picture *"Isti mirant stella"* reads, "They marvel at the star." In the adjoining scene a worried Harold, no doubt distressed by this portentous omen, is shown on his tottering throne (Fig. 7-2).

This comet caricature is the earliest contemporaneous representation of Halley's comet. It also pictures perhaps the comet's most famous appearance, for it not only coincided with the Norman invasion of England—it was also probably partially responsible for it.

Fig. 7-2. Halley's Comet of 1066 As Recorded on the Bayeux Tapestry. Pictured with the earliest contemporaneous representation of Halley's comet, a worried Harold II of England totters on his throne as a group of his countrymen gaze upon the omen in the sky.

During this pass, the comet's appearance was most remarkable. Joannes Zonaras, the twelfth-century Greek historian writing in his account of the reign of the Emperor Constantine Ducas (ca.1007-1067), described it as being as large as the full moon. The Belgian chronicler Sigebert of Gembloux (ca.1030-1112) noted that "over the island of Britain was seen a star of a wonderful bigness, to the train of which hung a fiery sword not unlike a dragon's tail; and out of the dragon's mouth issued two vast rays, whereof one reached as far as France, and the other, divided into seven lesser rays, stretched away towards Ireland."

At the time of the apparition during the spring of 1066, there were two competitors for the crown of England—Harold II and Duke William of Normandy. Harold Hardrada, King of Norway, was also about to launch a second front against the armies of Harold.

William of Malmesbury (ca.1095-1143), English chronicler and compiler of *Gesta Regum Anglorum (Acts of the Kings of the English)*, recorded how the apparition affected the mind of a fellow monk named Elmir of the monastery at Malmesbury. His country in danger of invasion in the north by the Norwegians and from across the English Channel by Duke William, Elmir lamented of the comet, "Thou art come, thou that will cause so many mothers to weep; I have seen thee long since, but thou seemest to me more terrible now that thou foretellest the ruin of my country."

While England viewed Halley's comet with foreboding, in Normandy Duke William regarded the comet as an auspicious sign. George F. Chambers in his *The Story of the Comets* recounts that, according to an ancient Norman Chronicle, astrologers had told William "how a star with three long tails appeared in the sky [and] that stars only appeared when a kingdom wanted a king, and how the said star was called a Comette." To the Duke, his destiny was clearly defined in the night sky.

Harold fought the Norwegians at Stamford Bridge on 25 September and defeated them. Then he hurried south with a tired army to meet William of Normandy. On 14 October the decisive Battle of Hastings was fought. By nightfall Harold lay dead and William the Conqueror had claimed the throne of England.

During the year 1145, Europe was preparing for the second Crusade, which began the following year. In the *Eadwine (Canterbury) Psalter*, a collection of the Psalms copied by the monk Eadwine from the earlier ninth-century *Utrecht Psalter*, a representation of Halley's comet appears at the bottom of the page with the text of Psalm 5 (Fig. 7-3). This drawing was probably made at or soon after the appearance of the comet.

When Halley's comet reappeared during the spring and summer months of 1456, all Europe was in a state of consternation. Three years previously, the Turkish army under command of Mohammed II (1430-1481), surnamed "The Conqueror," had captured Constantinople. Now he had crossed

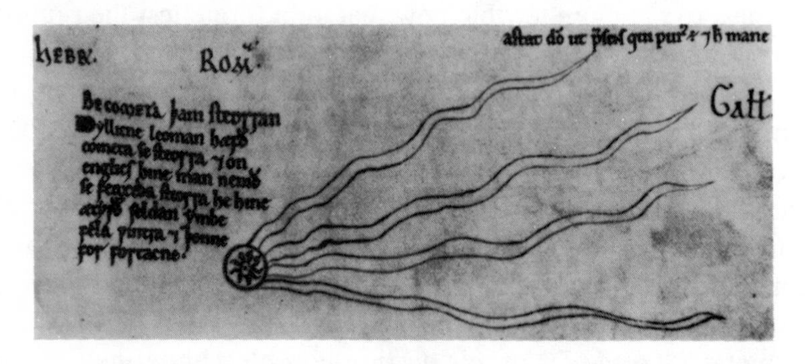

Fig. 7-3. Halley's Comet During 1145 As Represented in the *Eadwine Psalter*. This collection of the Psalms by the monk Eadwine contains a drawing of the comet, mentions its brilliance, and remarks that comets possess the ability to foretell events.

the Hellespont, was preparing to lay siege to Belgrade, and threatened to overrun all Europe.

Popular terror was heightened by the appearance of this brilliant comet, a magnificent celestial object with a tail extending more than a third of the way across the heavens. It was the precursor, it was assumed, of further calamities. To some eyes the comet's tail described the form of a Turkish scimitar and was regarded as a heavenly sign of the conflict raging between Christians and Infidels. A representation of this comet was later drawn by the Polish astronomer Stanislas Lubieniecky (1625-1675) in his *Theatrum Cometicum* (Fig. 7-4). The comet was at its brightest in early June and was seen gliding across the sky towards the moon. The fearful effect was intensified when a dark shadow hid the light of the moon from view—a lunar eclipse.

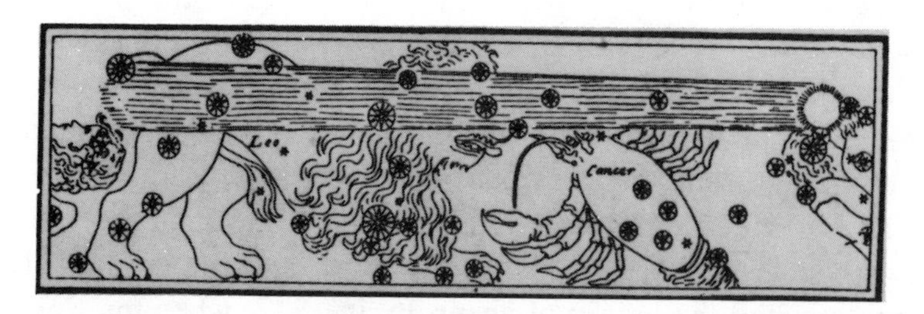

Fig. 7-4. Halley's Comet in 1456. The passage of the comet through the constellations of Leo and Cancer is shown by Stanislas Lubieniecky in his *Theatrum Cometicum.*

In *Comets* by Mary Proctor and A.C.D. Crommelin, an account of the comet's appearance by a Bavarian Jesuit named Pontanus is given. "Some persons seeing the darkness of the eclipse, and perceiving the comet in the form of a long sword advancing from the west and approaching the moon, thought that it presaged that the Christian inhabitants of the West would come to an agreement to march against the Turks, overcoming the enemy. While the Turks, on their part, taking into consideration the state of affairs, fell into no small fears and entered into serious arguments as to the Will of Allah." The prevailing opinion of the learned men of the time, however, was that the comet was an omen of divine displeasure and that a grievous calamity was about to befall the world.

The reigning pontiff was Calixtus III (1378-1458). He had been born Alfonso de Borgia, great uncle of the notorious Lucretia and Caesar Borgia. Shortly after the beginning of his pontificate in 1455, he issued a Crusade Bull, in which he urged a united action by the Christian leaders of Europe against their common enemy, the Turks. His plans met with little success, however, since the Wars of the Roses were beginning in

England, and the remainder of Europe was in a most unsettled state.

Controversy surrounds Calixtus's issuance of a second bull on 29 June 1456 when the comet still hung in the night sky above Rome. A legend gradually arose that the Pope was so alarmed by the apparition that he issued a bull against the comet, ordered bells to be rung to frighten it away, and commanded both prayers and processions to deliver the world from its influence. He also reputedly added to the *Ave Maria* the words, "Lord, save us from the devil, the Turk, and the comet." The myth reached the height of absurdity in 1832, when the French astronomer François Arago (1786-1853) suggested that the Pope "excommunicated at the same time the Comet and the Turks."

William F. Rigge, S.J. (1857-1927), Director of the Creighton University Observatory at Omaha, reviewed the circumstances surrounding the slander on the Church. In an article published in 1910, he notes that an investigation of the original documents reveals that the Papal Bull in question contains no reference to a comet, but does order the noon ringing of bells in every church throughout the world to remind the faithful to pray, a practice still continued today as the midday Angelus. The Bull also provides for processions to be held, the first of which took place on 4 July when indeed the comet was still visible, but there is no evidence that the processions are a direct result of the comet's appearance. The only contemporary source that connects the Pope with the comet in any way—and the probable source of the confusion—seems to be in the *Vitae Pontificum (Lives of the Popes)* by Platina (1421-1481).

Platina was commissioned in 1471 by Pope Sixtus IV to write a history of the popes. He completed the work in either late 1474 or early 1475, and it was published in Venice in 1479. Platina was a contemporary of Calixtus III and was living in Rome at the time the comet appeared. With respect to the Pope and the comet he writes:

A hairy and fiery comet having then made its appearance for several days, as the mathematicians declared that there would follow a grievous pestilence, dearth and some great calamity, Calixtus, to avert the wrath of God, ordered processions, that if evils were impending for the human race, He would turn all upon the Turks, the enemies of the Christian name. He likewise ordered, to move God by continual entreaty, that notice should be given by the bells to all the faithful, at midday, to aid by their prayers those engaged in battle with the Turk.

No reference is made of a papal bull, or an exorcism, or excommunication of either the comet or the Turks. Platina's competence as a historian has been questioned, and it is unclear whether he was correct in attributing the motive of the pontiff's orders to the fear aroused by the comet. It is possible that Platina combined a coincidental occurrence of events into a misrepresentation of history.

What is known for certain is that in July the forces of the Pope under John Huniades, Giovanni Capistrano, and the papal legate Carvajal repulsed Mohammed II and the Turks at Belgrade. But in 1521, with the Christian army decimated by an epidemic, Belgrade fell into the hands of the Turks.

In 1493, a representation of Halley's comet as it appeared in the year 684 was executed. This crude woodcut (Fig. 7-5) is to be found in the *Liber Chronicorum,* or *Weltchronik (World Chronicle),* of Hartmann Schedel, known in English as the *Nuremberg Chronicles* because it was printed in Nuremberg.

During the comet's next return in 1531, the Turks were still present in force on European soil, although their progress had been checked before the gates of Vienna. Unchecked, however, was the tide of the Protestant Reformation which continued its sweep across Europe.

Fig. 7-5. Representation of Halley's Comet in the *Nuremberg Chronicles*. This woodcut was made in 1493 and depicts the comet as it appeared during 684, a year when a series of calamities befell the earth.

When Halley's comet flashed upon the world in 1607, William Shakespeare (1564-1616) was still alive. Of his several allusions to the prophetic nature of comets, perhaps his most famous is Calpurnia's warning in *Julius Caesar:* "When beggars die, there are no comets seen; The heavens themselves blaze forth the death of princes." In May of that year, three frail ships bearing 105 colonists landed some fifteen miles inland from Chesapeake Bay at a site later to be named Jamestown, after the reigning Stuart king. This was the first permanent English settlement in America. And just as over 500 years before, Halley's comet again was present at the birth of a mighty nation.

Halley's comet was held by many Englishmen to be their "national comet." According to a quotation which appears in an article by Irene E. Toye Warner in 1909, the first predicted return of the comet in 1759 was "one of the most glorious years in the history of England, when it was necessary every morning to ask what new victory there was, for fear of missing one!" The English defeated the French in decisive naval battles; in India, English power gradually became supreme, and General Wolfe captured Quebec and Montreal and secured Canada for the British flag. This nationalistic fervor persisted through the most recent return. A quotation in the May 1910 issue of *Current Literature* notes how the comet "declares the successful work of an English astronomer, it recalls the past history of a glorious science, and reminds us how much of the progress is due to our countrymen."

One person who had a special interest in Halley's comet was Mark Twain. He was born on 30 November 1835 as the comet hung in the sky above Florida, Missouri. He wrote that he expected to die during its next visit, "The Almighty has said, no doubt: 'Now here are these two unaccountable freaks, they came in together, they must go out together.'" In the spring of 1910 he had "gradually let go of the world" following the deaths of his wife and daughter. As he lay gravely ill at his house called Stormfield at Redding, Connecticut, the comet came hurtling toward the sun once again. On 20 April, as one commentator wrote, "The mysterious messenger of his birth year, so long anticipated by him, appeared that night in the sky." Twain died the next day with the comet, according to Noel Grove, writing in the *National Geographic* of September 1975, "tracing a fiery tail across the night sky as clearly as he had illuminated the character of his countrymen."

The comet's return in 1910 also anticipated the end of a monarchy. King Edward VII of England died within a month following the appearance of the comet, events felt by many to be more than just a curious coincidence. The apparition also prompted fears of a German invasion of England. There were those who noted that both the years 1066 and 1910 contained a "10" and that the names of both past and potential aggressors of the island were similar: William the Conqueror and Kaiser Wilhelm.

And we can confidently predict that during the early months of 1986 there will be heard the voices of pseudo-scientists, psychics, and fanatics warning of the malign influence radiating from the cosmic interloper called Halley's comet.

CHAPTER 8

The Fearful Apparition

> **Through the glowing appendage of Halley's comet, as poisonous as it is beautiful, the earth will plunge.**
>
> **—Waldemar Kaempffert**

From the beginning of recorded history to the present day, all manner of fantastic and fearsome ideas have been associated with comets. The early observers of the heavens, ignorant as to the true nature of these apparitions, regarded them as the cause or at least as the prophets of war, famine, the downfall of empires, the death of kings, and universal distress for the inhabitants of the earth. Later, endowed with religious significance, comets became messengers of a wrathful God, harbingers of disaster placed in the skies by divine command. Since they were regarded as signs and portents, however, comets were believed to pose no direct physical peril to the earth.

When the comet predicted to return by Edmond Halley actually did so in 1759, it offered convincing evidence that comets were tangible members of the solar system rather than mystical messengers. This discovery helped to dispel the superstitious dread of comets, but it suggested a far grimmer possibility. As material objects moving through the cosmos at tremendous speed, they were capable of doing the earth significant physical harm. The old fear of a spirit of evil was replaced by a new one—that a comet would collide with the earth.

It was Halley himself who first considered the possibility that the earth might encounter a comet and thus come to a frightful end. According to the *Scientific American* of March 1910, Halley plotted the orbit of "a chariot of fire" which had alarmed many of his scientific contemporaries, including Newton's friend, William Whiston. His calculations led him to the startling conclusion that "the comet, when passing through the descending node, had approached the earth's path within a semi-diameter of the earth." After considering what the result of an actual collision would have been, he concluded that "if so large a body with so rapid a motion were to strike the earth—a thing by no means impossible—the shock might reduce this beautiful world to its original chaos." In fact, Halley believed it not improbable that at some remote period the earth had been struck by a comet, resulting in a change in the position of its axis of rotation.

In the year of the first predicted return of Halley's comet, John Wesley (1703-1791), the English evangelist and founder of Methodism, warned of the possible danger. "Were it to fix upon the earth...when it is some one thousand times hotter than a red-hot cannon ball, who does not see what must be the immediate consequence?" he asked.

Joseph Jérôme Lalande inadvertently created a panic in Paris in 1773 when he discussed the possibility of a collision between the earth and a comet in a paper entitled "Reflections on Those Comets Which Can Approach the Earth." Lalande noted that the orbit of a comet might be so perturbed by the planets that a collision with the earth might occur. This topic was intended for presentation before the Academy of Sciences, but it was not read because the rumor spread that Lalande had predicted the destruction of the earth by a comet on 20 May. Lalande was compelled to attempt to allay the public fears as well as he could in a soothing article published in the *Gazette de France*. In it he emphasized that the likelihood of a collision of the earth with a comet was extremely slight. Nevertheless, the people were not completely reassured. Designing priests sold places in Paradise at a very high rate and people begged the Archbishop of Paris to pray to God to alter the comet's course.

When Halley's comet returned in 1835, rumors again spread that it would collide with the earth. A *Natural History* article in December 1980 noted that Mennonites in America, believing the approach of the end of the world, allowed their crops to die on the vine. It was a time when "people were scared as wild geese...."

Fears of a cometary catastrophe persisted during the nineteenth century, promoted in part by the popular press and by writers of the day. In 1857 a French newspaper cartoon depicted a fiendish comet tearing the earth asunder beneath the amused gaze of the moon (Fig. 8-1). In 1877, Jules Verne, the French

Fig. 8-1. A Comet Strikes the Earth. In this 1857 French newspaper cartoon, an amused moon looks on as a comet collides with the earth, reducing it to "its original chaos."

novelist and scientific prophet, wrote *Hector Servadac* in which the earth is smashed to bits by a comet. In the 1906 novelette, *In the Days of the Comet,* by H.G. Wells, the earth is enveloped in a mysterious gas from a comet's tail.

Some remnants of comet fear still persisted when Halley's comet returned to the earth's vicinity in 1910. No major comet had appeared for several decades, so that most people had never seen a comet before. The intemperate comments of a handful of scientists who should have known better were disseminated worldwide by a communications system that had made recent rapid advances. These reports, combined with the usual apprehension associated with unusual celestial events, led to a fear which reached nearly panic proportions.

As the comet headed out from the sun following its closest approach, an unusual event was scheduled to occur. The earth, the head of the comet, and the sun would be in *syzygy,* that is, in direct line with each other. And on this date, 18 May, the earth would pass through the comet's tail (Fig. 8-2). Twice before, in 1819 and 1861, the earth had been swept by a comet's tail and no one was the wiser until long after. However, this passage was different.

In January 1910, Professor Edwin B. Frost, Director of the Yerkes Observatory, made a disturbing discovery. His spectroscope had detected the probable presence of cyanogen gas in the tail of Halley's comet (Fig. 8-3). Cyanogen is both inflammable and highly poisonous and forms the basis of cyanide compounds. One astronomer, according to the *New York Times,* was "of the opinion that the cyanogen gas would impregnate the atmosphere and possibly snuff out all life on the planet." Most other astronomers disagreed, but a widespread feeling of uneasiness remained.

Camille Flammarion (1842-1925), an imaginative French popularizer of astronomy, described in his "La Fin du Monde" the possible effect of passing through the tail. "The comet's tail is composed of deadly cyanogen and other gases, including

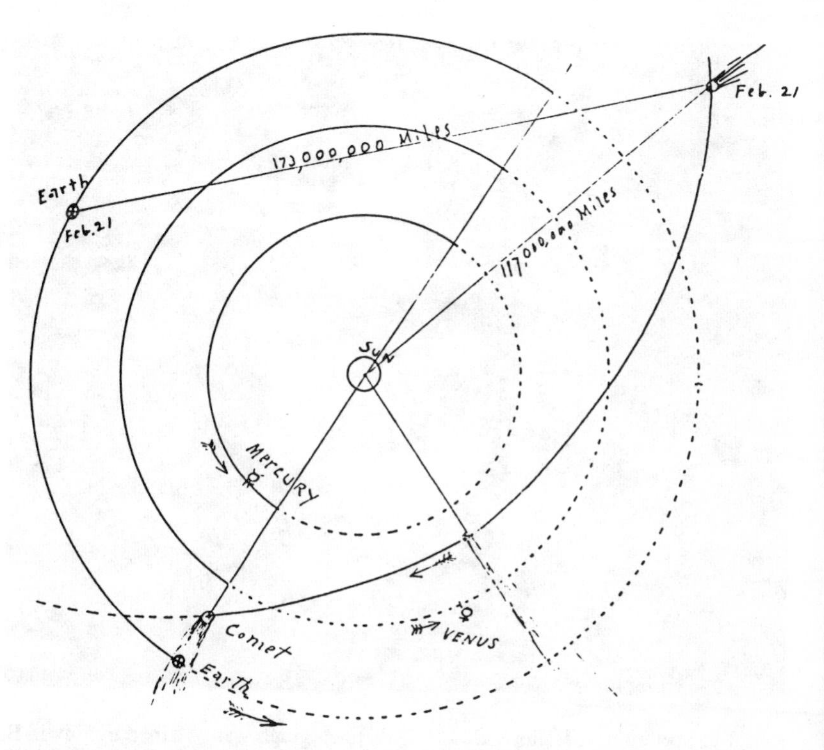

Fig. 8-2. Syzygy of the Sun, Halley's Comet and the Earth. This diagram by the American astronomer Edward Emerson Barnard shows the alignment of the three astronomical bodies on 18 May 1910, when the earth passed through the comet's tail.

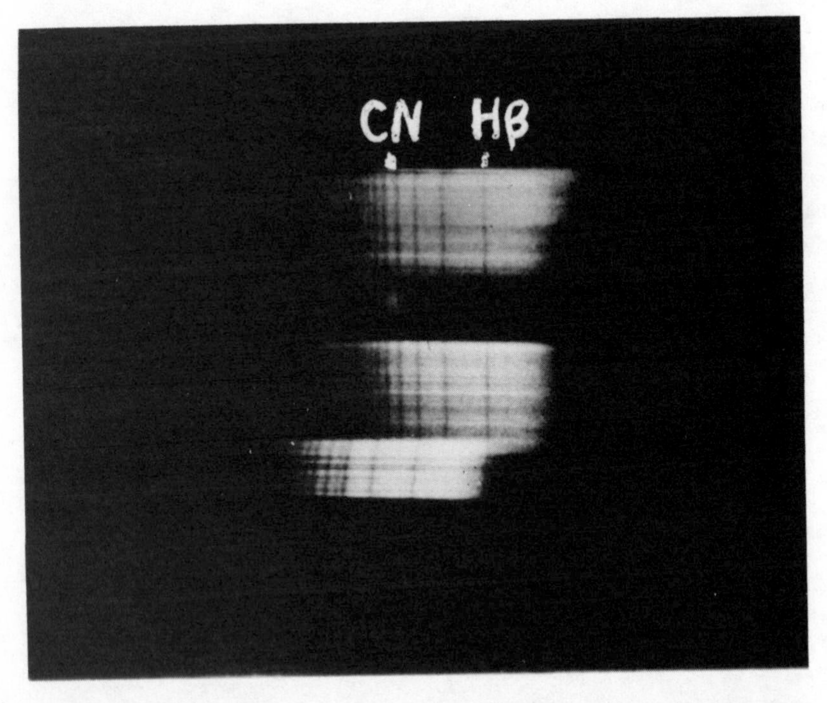

Fig. 8-3. Spectrum of Halley's Comet. This photograph was obtained by Edwin B. Frost and John A. Parkhurst of Yerkes Observatory on 14 January 1910. The comet's spectrum lies between the two very broad and conspicuous "comparison spectra" of the star lambda Tauri. The bright line marked "CN" is the so-called "cyanogen band" which may be produced by poisonous cyanogen gas.

hydrogen. If the earth should pass through this tail, either the hydrogen will ignite, blasting earth asunder in a gigantic explosion, or the comet gases will sweep aside our own atmosphere, reacting with the nitrogen to form the familiar laughing gas, nitrous oxide, and suffocating all animal life in a ghastly parody of death."

The thought that quite soon a deliriously happy mankind would dance itself to an anesthetic death did not enchant most people. Their fears were reinforced when Professor Henri Deslandres of the Astrophysical Observatory at Meudon gave a tentative endorsement to Flammarion's theory in early May. "The hypothesis that the gas (in the comet's tail) is liable to affect terrestrial atmosphere would not be at all absurd."

As Bill Stephenson was later to write in "The Panic Over Halley's Comet":

> Within days these two statements, seized upon by delighted newsmen and grossly exaggerated, had been flashed by wireless across the world. Everyone who heard the story had only to search his own sky to find verification: a glowing dot brighter than most stars, though not nearly so bright as the moon. Soon its tail was visible, that dread tail that would snuff out all life on earth... The comet had never come this close to earth before, the story said. What if it should smash right into the earth, and that terrible tail swept into every home and field, suffocating men, women and children without mercy?

Other scientists added to the alarums. At the Paris Observatory, an astronomer named Marchand declared that variations in Halley's path could conceivably "bring about unexpected results." D.J. McAdam, Professor of Astronomy at Washington and Jefferson College, noted the vagaries of comets and asked of other astronomers, "How can they tell what a comet is going to do, when it doesn't know itself?" He proposed that a direct collision with the nucleus of the comet was not beyond the realm of possibility since "the comet will do what it has never done before, and the probabilities were millions to one against its doing. It will pass directly between the earth and the sun, transit the sun's face, and envelop the earth in its tail." And of that tail he wrote, "Disease and death have frequently been ascribed to the admixture of cometary gases with the air. Enough of such gases as are in the comet's tail would be deleterious."

The fears of the late British astronomer Richard A. Proctor about a startling form of danger were resurrected. "The intermixture of cometic matter with the atmosphere of our earth might not always be a perfectly innocuous process," he had said. "Suppose that a comet composed in the main of hydrogen should mix with our air until the oxygen of the air and hydrogen of the comet were in the proportion in which they are present (chemically combined) in water. Then, unless every fire and light in the whole world were extinguished, there would be a tremendous explosion, followed instantly by a deluge of water, and leaving the burnt and drenched earth no other atmosphere than the nitrogen now present in the air, together with a relatively small quantity of deleterious vapors."

There was a self-styled prophet by the name of Lee Spangler, who had purportedly predicted to the day the San Francisco earthquake, the death of Queen Victoria, and the assassination of President McKinley. He discounted predictions that the earth would pass through the tail of the comet; instead, he said, the head of the comet would collide with the earth!

Of this fearful prospect Waldemar Kaempffert, a popular science writer of the time, speculated on the consequences:

> This globe would be punctured like a bubble, and all the molten rock, the steam, and the gases so long pent up within the thin shell on which we live would spurt forth in a white-hot deluge. Mountains would topple; continents would crumble like glass; rivers and oceans would vaporize into clouds of hissing steam. Out of the maelstrom of lava and debris the earth would emerge, a

smoldering, planetary ember, lifeless, but still glowing with the heat of a mighty cataclysm.

And finally, there was the additional worry that if a brush with the comet should displace the magnetic pole, the earth's electrical potential would be altered and every inhabitant of the globe would be electrocuted.

The feeling of vague apprehension was not restricted just to the ignorant and unschooled portion of the populace. Impressed by the fact that no scientific authority had actually denied that the earth would probably pass through the comet's tail, educated people, too, recorded the comet's progress with alarm. In London, *The Times* of 13 May noted that "even the cultured people of France are said to await the comet's approach with dread." The particular fears of each segment of the population were spread widely by daily newspaper accounts, with each group's apprehensions feeding those of the others.

In an attempt to allay these anxieties, sensible and reasoned scientific voices were raised. There would be no collision, they said, and furthermore the gases within the tail were too tenuous to do anyone any harm. Professor Leon Campbell of the Harvard Observatory announced that "there is absolutely no chance for harm to come to anyone," but he did suspect that a luminous display was possible. The Naval Observatory in Washington, D.C., issued a terse statement that "no disturbances" would occur. Dr. S.A. Mitchell, Professor of Astronomy at Columbia, said that he was "...almost absolutely certain that no harmful effects will come to us." An article in the *Scientific American* of March 1910 noted that "...the wild tales of the possible effects of poisonous gases, tales for which the newspapers are very largely responsible, are utterly without foundation." But for many individuals the time for such assurances was past.

Following the first visual sighting of the comet in Curaçao on 20 April, reports of increasing consternation among people throughout the world began to appear in the press. In the country churches and monasteries of Southern Russia, prayers were offered for the salvation of the motherland from the threatened cataclysm. A message to the *Frankfurter Zeitung* from Budapest reported the suicide of a landowner by the name of Adam Toma. He said that he preferred death by his own hand to being "killed by a star."

In Turin, Italy, so great was the alarm that the *Gazeta,* in an attempt to allay the fear, published an article "pledging the honor" of its editors that no harm would come from the celestial visitor to the earth or its inhabitants.

In Port Au Prince, Haiti, the superstitious fears of the natives were exploited by a shrewd old Voodoo doctor who was selling "comet pills" as fast as he could make them. In South Texas, similar medicinals containing a combination of sugar and quinine, which were promoted as enabling the body to withstand the gases of the comet's tail, sold like hotcakes.

From Walla Walla, Washington, came a report of a French sheepherder driven insane from worry over the appearance of Halley's comet. The Frenchman had read all the literature possible on the comet and while watching his flocks on the banks of the Snake River tried to fathom its nature and purpose. As a result of his ponderings, he ended up confined in a padded cell in the county jail.

From Denver came notice of more tragedy. A young woman was driven to despair by brooding over the possibility of the annihilation of the world by the comet. She attempted suicide by swallowing morphine. Reportedly her last words after taking the poison were, "I think the comet..."

In the Midwest lightning rods were removed from the roofs of barns and houses in case they would attract the comet, and some lowland residents near Lake Superior fled their homes, fearing a tidal wave. In Mexican towns and villages, candlelight street processions were held. Pennsylvania coal miners refused to enter the pits, wishing to be on the surface when the world ended, and Indian tribes in western Canada performed ritualistic dances to protect themselves from the menace which they saw in the night sky.

Of course not all people viewed the comet with apprehension. Most believed the majority of scientists when they said that no harm would come to the earth. These people understandably adopted a more frivolous attitude toward the comet.

In Chicago a newborn baby was named for the comet. The doctor present at her birth was given permission to name her. "You henceforth will be known as 'The Comet Girl,' " he said, "and I choose for you the name of 'Halley.' "

In Istanbul, Turkey, the police took the heaven-sent opportunity presented by the entire population's crowding onto the rooftops to pray each night to round up ferocious dogs that the citizens had always kept them from capturing.

New York City had its own unique manner of welcoming the comet. New drinks were concocted and named "Cyanogen Cocktail" and "Syzygy Fizz." "Comet parties" were held on the rooftops of many of the hotels, where people spent the greater part of the night gazing skyward through telescopes and opera glasses while eating late-night meals (Fig. 8-4). On the Hotel Gotham roof garden, small silver telescopes were provided as favors and an orchestra played "A Trip to Mars." Riverside Drive was crowded with sky-watchers all night long as the comet approached at forty-three miles a second.

In Paris all the big cafes and restaurants scheduled special "comet suppers" and a financial success was expected. In Toronto the jewelry firm of Ryrie Brothers announced the sale of "Halley's comet Jewelry" which their advertisement described as "tie pins and brooches set with opals, moonstones or peridots. Peridot is of meteoric origin and is the fashionable stone this season."

Finally that awful day—18 May—arrived when the comet would pass between the earth and the sun, change from the morning to the evening sky and bathe the earth with its tail. It

Fig. 8-4. A "Comet Party." This particular get-together was held on the roof of a New York City hotel. Such "parties" reflected the intense interest in the brilliant apparition of Halley's comet in 1910.

was also near its closest approach to the earth, about 14 million miles distant. Popular interest and hysteria reached its peak. But so did scientific interest. Careful and costly preparations had been made to note any minute effects upon the earth that might occur as a result of this passage through the tail. All unusual accounts of atmospheric phenomena were to be recorded.

The tail first brushed the earth during the early evening in the United States. It would remain in contact with the earth for several hours. As the world held its breath, all eyes were directed skyward. Some auroral displays and iridescent clouds were seen, but the air remained breathable and the earth unscathed. Nothing else of consequence was reported, except that the tail itself was lost to view, prompting the opinion in the *Montreal Gazette* that "earths must be dangerous to comets."

Fig. 8-5. An Imaginative Rendering of the Visit of Halley's Comet in 1910. For some anxious observers, however, this exaggerated appearance reflected their impressions of the event.

Several days later the tail reappeared in the evening sky, an object of beauty and wonderment, no longer evoking terror and consternation—at least not until its next encounter with our planet.

CHAPTER 9

The Nature of the Object

Comets are the nearest thing to nothing that anything can be and still be something.

—National Geographic Society press release, 31 March 1955

Our fascination with comets can be attributed as much to their mystery as to their beauty. They appear suddenly in the sky. Their origin is unknown. Their visibility is tantalizingly brief. Their composition is a matter of conjecture. Despite the fact that Halley's comet has been observed throughout recorded history and has figured significantly in our perception of the universe, it and its celestial kin are probably less understood than any other members of our solar system. The infrequent visitations and short viewing opportunities have contributed to our relative ignorance about comets. Since the middle of this century, however, new ideas and observations have combined to produce a modern understanding of comets which has shed some light on their mysterious nature.

These celestial wanderers are classified as either *short-period* or *long-period* comets, depending upon the time it takes them to complete one orbit. Of the comets so far observed, approximately one in every five has a period of less than two centuries; this earns it the designation of short-period comet, or simply "periodic" comet. These number about 100. All others are termed long-period comets and include most, if not all, of the comets with parabolic or hyperbolic orbits. The long-period comets may shine brilliantly and be easily seen with the naked eye. Compared with these, the short-period comets are generally small, faint objects seldom visualized without the aid of a telescope. A notable exception is Halley's comet, which is both large and bright.

The names of the short-period comets are prefixed by the symbol "P/" of which P/Halley is the best known example. The period of Comet Halley is approximately 77 years, which places it about in the middle of the short-period comets whose periods range from Comet P/Encke of 3.3 years to Comet P/Herschel-Rigollet of about 155 years.

It was Pierre Laplace who initially determined that the orbits of certain comets were changed from parabolas to ellipses by the gravitational attraction of the massive outer planets. During the close encounter of the comet with a large planet, especially Jupiter, the velocity of the smaller body may be retarded to such an extent that its orbit is changed into an ellipse. The comet is in effect "captured" by the planet, and the *aphelion,* or point of the comet's orbit most distant from the sun, now corresponds closely to that planet's orbit (Fig. 9-1). As a result of the capturing effect, we can refer to "families" of comets orbiting about Jupiter, Saturn, Uranus, and Neptune. The Neptune family numbers Halley's comet among its members.

The gross structure of a comet is composed of four distinct parts (Fig. 9-2). The *nucleus,* usually no more than a few miles in diameter, is the central substance from which the other parts evolve. The *coma* is the hazy envelope of sublimated gas and dust that obscures the nucleus from direct view and creates the nebulous outline associated with comets. The *tail* is the most identifiable cometary feature, stretching for great distances in an antisolar direction. The fourth component, an immense *hydrogen cloud* that surrounds the comet and extends for

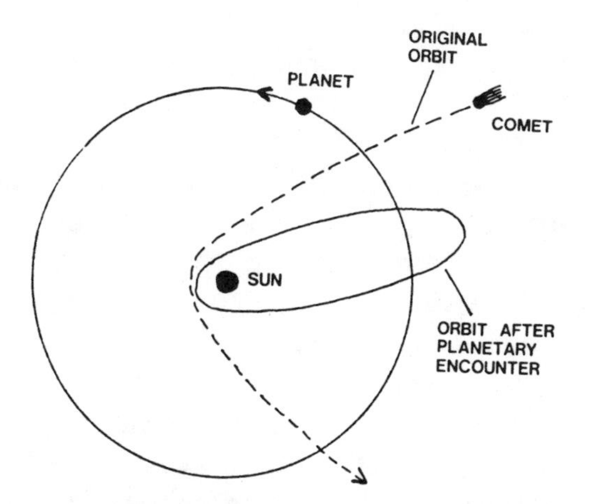

Fig. 9-1. The "Capture" of a Comet by a Planet. It is postulated that the planet Neptune converted the orbital path of Halley's comet into an ellipse by this process.

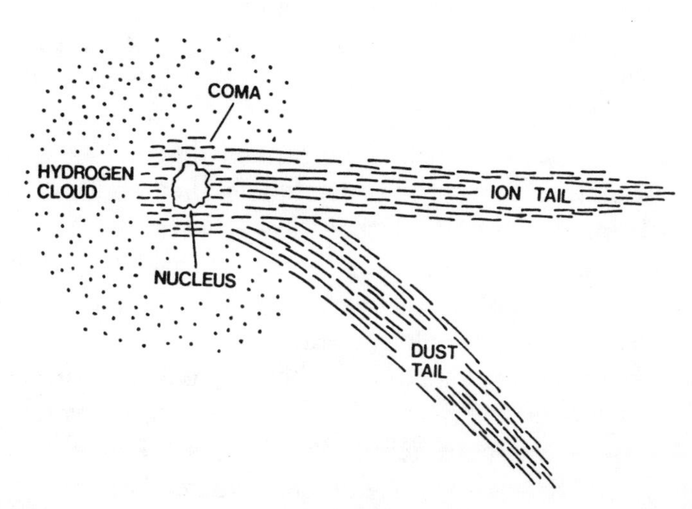

Fig. 9-2. The Structure of a Comet. The nucleus and the coma together form the comet's "head." The hydrogen cloud is invisible to the naked eye. The ion tail is straight and the dust tail is slightly curved.

millions of miles, was discovered only recently in 1970. The ultraviolet spectrometers of NASA's Orbiting Astronomical Observatory (OAO-2) and Orbiting Geophysical Observatory (OGO-5) detected the large hydrogen cloud surrounding Comets Tago-Sato-Kosaka and Bennett. Comet Kohoutek, the celebrated "Flop of the Century" because of its disappointing visual show, was later observed also to possess such a cloud. Invisible to the naked eye, this cometary hydrogen cloud is apparent only in the very far ultraviolet portion of the electromagnetic spectrum. These observations verified a hypothesis made in 1960 by Ludwig F. Biermann of the Max Planck Institute for Physics in Göttingen that if frozen water is assumed to be the major component of the nucleus, then a huge cloud of hydrogen should surround the comet.

The volume of a comet greatly exceeds that of any other member of our solar system including the sun. Yet its density is so low that stars shine undiminished through the tenuous tail, and its nucleus has never produced significant perturbative effects upon the motion of any planet or satellite. On 18 May 1910, when Halley's comet transited the disc of the sun, no solid nucleus could be observed and an upper limit of thirty miles was placed upon its size.

Exactly how the comets were formed remains unknown. One speculative theory considers them the remnants of a disrupted planet. In 1930, the Soviet astronomer S.K. Vsekhsvyatsky advanced the idea, as had the French mathematician Joseph Louis Lagrange before him, that comets are the ejecta of violent volcanic eruptions on the surface of the major planets or their satellites. Although we now know that a satellite of Jupiter, the moon Io, does possess active volcanos, this theory is not considered likely. In 1948, the British astrophysicist Raymond A. Lyttleton attributed both the origin and capture of comets to the accretion of matter as the sun with its gravitational influence passed through interstellar dust clouds.

Today most scientists believe that the comets were formed within the solar system itself, and that they originated in the swirling cloud of dust and gases of the original solar nebula, that great fog of matter which later condensed to form the sun and the planets. If this is true, the comets would be primordial chunks of matter that have remained unchanged since the birth of the solar system. As quoted in *Sky and Telescope* for February 1981, a NASA Comet Science Working Group report has stated that comets "are probably the most primitive bodies remaining in our solar system and may preserve the clearest record of the chemical and physical processes which marked the beginning of the solar system 4.6 billion years ago." A close study of these ancient artifacts might provide some answers to questions we have about the formation of the earth and the rest of the solar system.

The Harvard astronomer, Fred L. Whipple, a noted personality of American comet studies, wrote in 1976 that "three explicit new ideas...revolutionised our concepts of comets." These theories were developed in the early 1950s and were, in order of publication: (1) Jan Oort's hypothesis that there is a great comet cloud about the sun; (2) Whipple's suggestion that a comet's nucleus is similar to that of a "dirty snowball"; and (3) Ludwig Biermann's concept of the relationship between the solar wind and the ion tails of comets.

If the comets are in fact true members of our solar system and not cosmic intruders, then their source becomes an item of intense curiosity. In 1932, the Estonian astronomer Ernst Öpik calculated that the sun's gravitational attraction would exert an influence over a cloud of meteors or comets to a distance that extends out almost as far as the nearest star, without the loss of a large fraction of them even during billions of years. In 1950, the Dutch astronomer Jan Hendrik Oort, director of the Leiden Observatory, postulated the existence of a huge cloud of comets, now known as "Oort's cloud," which contains as many as 100 billion comets in a spherical shell in orbit about the sun and under its gravitational influence, but extending trillions of miles from it. He determined that the gravitational attraction of an occasional passing star would disturb the orbits of these comets and deflect some into the inner planetary region where they would be influenced by the major planets and become either long-period or short-period comets.

Any model of the physical structure of comets must satisfy the wealth of observational data obtained over several centuries. As a comet approaches the inner planets on its sunward journey, it gradually becomes active, manifesting its presence by the development of a coma and tail. The invention of the spectroscope in the late nineteenth century aided in defining the chemical composition of the head and tail of comets. Both gas and dusty material were found. The gas consisted of *radicals,* that is, fragments of parent molecules such as methane, ammonia, and water torn apart by the action of the ultraviolet in the sun's radiation, plus other non-ionized material. Although the spectrum of a comet reveals information about its internal chemistry, any detailed description of its physical structure could only be inferred from the comet's observed behavior.

Giovanni Schiaparelli (1835-1910) was an Italian astronomer who extensively studied the planet Mars. He named the straight features that he observed on the planet's surface *canali* or "channels." A mistranslation of his word as "canals" helped prompt speculation about intelligent life on Mars. Schiaparelli also investigated comets, and in 1867 he demonstrated a connection between the periodic comet Swift-Tuttle and the annual Perseid meteor shower.

As a comet orbits about the sun, solid material is continually expelled in its path. A small fraction of a comet's mass is lost during each orbit. When this debris enters the earth's atmosphere, the result is a shower of "shooting stars." A cometary origin for recurring meteor streams is now accepted, and at least seventeen major permanent meteor streams and a smaller number of temporary meteor streams have been observed to intersect the earth's orbit.

Halley's comet is associated with two meteor showers each year. For several days in early May, the Eta Aquarid meteor shower can be seen. It is so named because the swarm radiates from the direction of the star Eta in the constellation of Aquarius. In October, the Orionid meteor shower occurs. Following the passage of Halley's comet in 1986, an increase in the number of meteorites observed in these two showers can be expected.

The cometary debris that does not fall upon the earth, the sun, or the other planets is scattered throughout interplanetary space. It is believed that as a source of small particles, comets contribute significantly to formation of the *zodiacal light*. This phenomenon has been observed for ages and in the Orient is called the "fox's tail" or "false dawn." It is best seen during the early morning in the eastern sky or as an afterglow of white light in the western sky just after sunset. Under very favorable viewing conditions, it also appears as *gegenschein*, or counterglow, a faint luminous patch in the night sky in a direction opposite that of the sun. The diffraction and reflection of sunlight by the cloud of fine dust near the earth's orbit produces this brightening of the night sky. In 1914, the Russian astrophysicist Vasilii Fesenkov was the first to suggest a cometary origin for such dust.

A peculiarity of comets is their unexpected and wayward motions, which seem to defy gravity itself and are the reason comets often do not appear on time. The law of gravitation formulated by Newton made posible the prediction by Halley of his comet's subsequent return. Yet the comets in their motion seem to contradict this very law. They speed up or slow down unpredictably. Significant systematic discrepancies between observed and computed positions of the periodic comets sometimes remain, even after all gravitational interactions have been properly allowed for. The existence of some sort of nongravitational motion in comets was initially inferred from our inability to predict accurately the return of short-period comets.

The most thoroughly studied comet is probably Comet Encke, since its period of revolution around the sun is a mere 3.3 years. This comet was first discovered in 1786 by Pierre Méchain (1744-1804). It was named for the German mathematician and physicist Johann Franz Encke, who computed the orbit and period of the comet. In 1819, Encke studied the motion of the comet and concluded that it deviated "wildly" from predictions based on Newton's law of gravitation. He found that the comet returned two and a half hours earlier with the completion of each revolution. In 1823, Encke postulated that the decrease in the period was due to a "resisting medium" in space, which caused a reduction in the dimensions of the comet's orbit. Once this notion was abandoned, subsequent explanations suggested that the problem was imperfect calculations or the failure to take all the planetary influences, including the minor planets, into consideration.

In 1836, Friedrich Bessel (1784-1846), the German astronomer and mathematician, suggested the possibility that nongravitational forces were at work on comets. In 1845, Herrmann Westphalen of Hamburg, following a suggestion from Bessel, tried to prove the existence of nongravitational forces acting upon Halley's comet when it returned in 1835. He thought that these forces might somehow be related to either the ejection of the comet's tail or to the solar atmospheric drag upon the comet. His results, however, based upon a total of 311 observations obtained over a ten-month period, failed to demonstrate that any systematic effect was acting upon the comet.

Historically, the prediction of a forthcoming apparition of Halley's comet has never been completely successful. The calculation by Cowell and Crommelin of the perihelion passage in 1910 was more precise than any previous ones, for they had included perturbations in the comet's orbit caused by six of the planets, including Neptune. But despite their great care, there remained a difference of 3.03 days between observation and calculation. There seemed to be no explanation, since this was greater than the error due to possible deficiencies in their computations.

Unlike Comet Encke which kept arriving early, Comet Halley was persistantly late. Other comets such as Biela and d'Arrest were also observed to be late in reaching perihelion. We now know that nearly all the orbits of short-period comets demonstrate *secular perturbations*, that is, changes that occur over very long intervals of time, which are the result of nongravitational forces at work upon them. In 1972, Tao Kiang of the Dunsink Observatory in Ireland determined that during its past eleven apparitions, Halley's comet persists in arriving late by an average of 4.1 days. This secular term must therefore be incorporated into any mathematically based predictions of its date of perihelion passage.

Astronomers have sought to explain this varied behavior of the comets by formulating a hypothetical model of the comet's nucleus. In the late nineteenth century, Richard Anthony Proctor (1837-1888), a British born astronomer, likened the nucleus to a sand bank or gravel bank formed mostly of dust and small pebbles, with gases trapped in the spaces between the particles. The absorbed gases would then be released as the comet approached the sun. Over a half-century later, another Briton, Raymond A. Lyttleton championed this "flying gravel bed" or "flying sandbank" theory. He envisioned the nucleus as nothing more than a mass of dust or small solid particles, held loosely together by gravity.

It remained for Fred Whipple, however, in a series of articles published in the *Astrophysical Journal* during the early 1950s, to postulate a nuclear model which best fit the observed physical behavior of a comet. His model consisted of a conglomerate of ices coarsely mixed with meteoric materials—an "icy-conglomerate" or "dirty snowball." As these frozen clumps of gas and dust left over from the formation of the solar system approached the sun, the action of solar heat would cause remarkable changes. The frozen gases would become heated and sublime or vaporize. These vapors would then, Whipple suggested, carry some of the dust particles with them and as they proceeded away from the nucleus would create the comet's atmosphere (coma) and tail phenomena.

Whipple's model also explained the dynamical behavior of the comets. The emission of gases on the sunward side of the nucleus would create a "jet effect." This rocket-type reaction from a rotating nucleus would either speed up or slow down the comet, depending on the direction of the jet.

These nongravitational jet forces suggested that comets possess solid nuclei. Radar echoes from the nucleus of Comet Encke in November 1980 and from Comet Grigg-Skjellerup in May 1982 support the hypothesis of a solid icy-conglomerate nucleus.

As a comet nears the sun, it acquires its most characteristic feature when the gases and dust are swept back to form one or more glowing tails. The shape and number of such tails vary greatly among comets, and changes in an individual tail over brief intervals of time also occur.

The Russian astronomer Fedor Bredichin (1831-1904) developed a classification of comet tails based upon their composition, length, and curvature. He distinguished the ionized gas tail which is generally narrow and very straight from the broader and brighter pure dust tail which was curved like a scimitar. The antisolar direction of the comet tails suggested a repulsive force emanating from the sun. It was first believed that the pressure of the sun's light provided this force. Proof of the existence of this radiation pressure was shown in 1901 by the Russian physicist Pëtr Lebedev (1866-1912). He demonstrated that extremely small pressure is exerted by light and was able to measure it. This finding satisfactorily accounted for the appearance and shape of dust tails, but it did not explain the tremendous velocity exhibited by ionized tails. An explanation for this only came with the discovery of the *solar wind* in 1951.

The sun steadily emits a stream of charged particles into space. This corpuscular radiation, consisting of electrons and positive ions, moves at speeds of up to 1,000 miles per second. When these particles strike a comet, they create a narrow tail of ionized gas several million miles in length pointing directly away from the sun. It was Ludwig F. Biermann who first associated this wind of charged particles with the high accelerations observed in the ion tails of comets.

Many of the most fundamental questions about comets, however, are unlikely ever to be answered by ground-based or near-earth observations alone. A determination of the structure, composition, and activity of the cometary nucleus; the chemical and physical processes occurring in the coma; and the nature of the interaction of the comet with the solar wind must all await the direct interception of a comet by an appropriately instrumented spacecraft.

Halley's comet is the logical target for such a space mission. It is a large, bright comet, and its orbit is very accurately known. Even in the case of Halley's comet, however, such a mission is not an easy matter. The inclination of Halley's orbit means the comet will have to be intercepted as it passes through the ecliptic. Moreover, because it circles the sun in a retrograde orbit opposite to that of the earth and the other planets, a quick flyby is the most practical and economic alternative to a more optimal rendezvous mission. Still, this once in a lifetime opportunity to study such a historic comet must not be missed.

Several robot missions are scheduled by the Russians, Europeans, and Japanese to fly past Halley's comet in early 1986. Because of budget restrictions, a planned United States mission to the comet was cancelled. Instead, NASA has diverted a spacecraft, which has been in orbit since August 1978 and was designed primarily to monitor the sun, onto a course to intercept Comet Giacobini-Zinner. Renamed the *International Cometary*

Explorer, the space probe will pass through the tail of this comet on 11 Sept 1985. Should it survive this encounter, the 1,054-pound ship would then be available to obtain data from Halley's comet at a distance of about 2 million miles.

The most ambitious space mission is being prepared by the European Space Agency with its *Giotto* spacecraft, named in honor of the fourteenth-century Florentine artist, Giotto di Bondone, who depicted Halley's comet in one of his frescoes. This mission is designed to intercept Halley's comet during the post-perihelion portion of its orbit. The spacecraft possesses imaging capability and will carry a 130-pound science package to within several hundred miles of the comet.

In addition to these observations from space, a massive worldwide ground-based effort known as the International Halley Watch has been formed which will involve hundreds of professional and thousands of amateur astronomers. Their combined observations, using every modern technique available, will help to produce the most complete body of data ever assembled about a comet.

The knowledge accumulated as a result of this intense scrutiny of Comet Halley will finally enable us to fulfill the prophecy of Lucius Annaeus Seneca who, when writing about comets in his *Quaestiones Naturales* in about the year A.D. 64, said, "The time will come when those things which are now hidden shall be brought to light by time and persevering diligence. Our posterity will wonder that we should be ignorant of what is so obvious."

CHAPTER 10

Epitaph for a Comet

Comets by the discharge of their tails are used up and eventually fall prey to Death.

—Johannes Kepler, 1619

Early on the morning of 16 October 1982, two astronomers, David C. Jewitt and G. Edward Danielson, both of the California Institute of Technology, were using the 200-inch Hale telescope at Palomar Observatory near San Diego in a search for the returning Halley's comet. For five years, astronomers had been scanning the skies in an unsuccessful attempt to be the first to sight the comet. In place of a photographic plate which captured the comet's image during its previous return, the two men used a highly sensitive silicone-chip light detector called a charge-coupled device, which is about fifty times as sensitive to light as photographic film. Aided by the ephemeris calculated by Donald K. Yeomans of Caltech's Jet Propulsion Laboratory, they pointed the giant telescope in the direction of the constellation Canis Minor. By making a series of six-minute exposures, they detected a faint moving dot against a background of stationary stars. Lying beyond the orbit of Saturn, Halley's comet was detected more than a billion miles away (Fig. 10-1). The comet was found only 8 arc seconds west

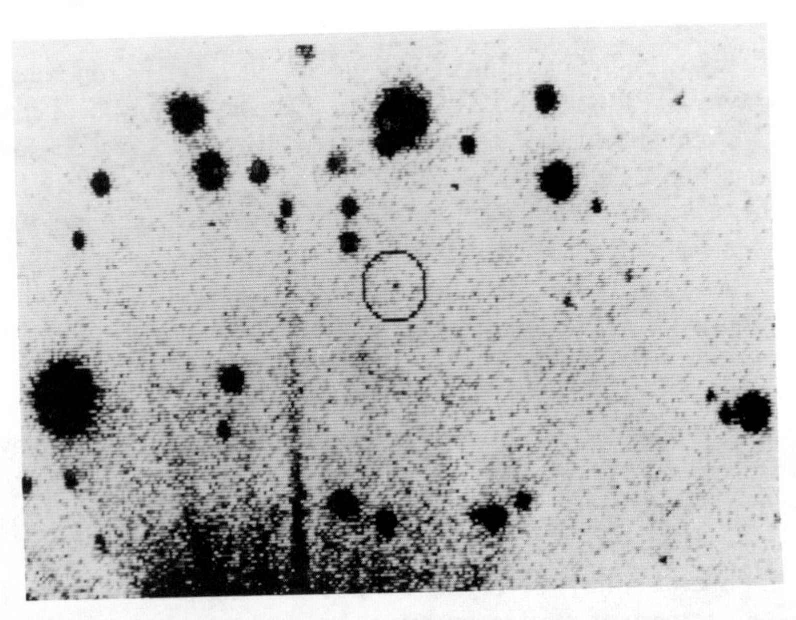

Fig. 10-1. Recovery Photograph of Halley's Comet. This is the famous comet (circled) as it appeared to astronomers on 16 October 1982. An advanced electronic detector system and the 200-inch Hale telescope at Palomar Observatory were used to obtain this photograph from a distance of more than a billion miles.

of the predicted position and was, according to Yeomans, "as faint as the light from a single candle seen 27,000 miles away."

As Halley's comet nears the sun, it will not appear as spectacular to the unaided eye as it did in 1910. Because of the unfavorable positions of the comet, sun, and earth (Fig. 10-2), this apparition may disappoint much of the general public, as did Comet Kohoutek during January 1974. Nevertheless, to those observers without high expectations, Halley's comet will prove to be a rewarding sight.

During the pre-perihelion portion of the comet's orbit, the best observing prospects will be in the evening sky of the Northern Hemisphere. The beginnings of a faint tail may be seen with the naked eye toward the end of December 1985 and during the first three weeks of January 1986, prior to the comet being lost in the evening twilight. Following perihelion, Halley's comet will reappear in the morning sky when the Southern Hemisphere will be favored for viewing. The tail may reach up to 40 degrees in length, and the comet should appear most brilliant when it passes closest to the earth on 11 April at a distance of approximately 39 million miles. Toward the end of April, the comet, again an evening object, will gradually fade from sight. The maximum brightness of Halley's comet during this return (magnitude 3 or 4) will be equivalent to that of Comet Kohoutek when that comet appeared in a similar position in the southern sky twelve years earlier.

Although the coming apparition is now virtually assured, future returns are not so certain. The lifetime of a short-period comet such as Halley's is relatively brief, perhaps only a few thousand years, and comets in general have a way of doing the unexpected. Based on our knowledge of past cometary events and considering the natural evolution of comets, several scenarios for the ultimate fate of Halley's comet can be postulated.

The present orbit of Halley's comet, in actuality a summation of perturbative effects, is the result of its prior interactions with

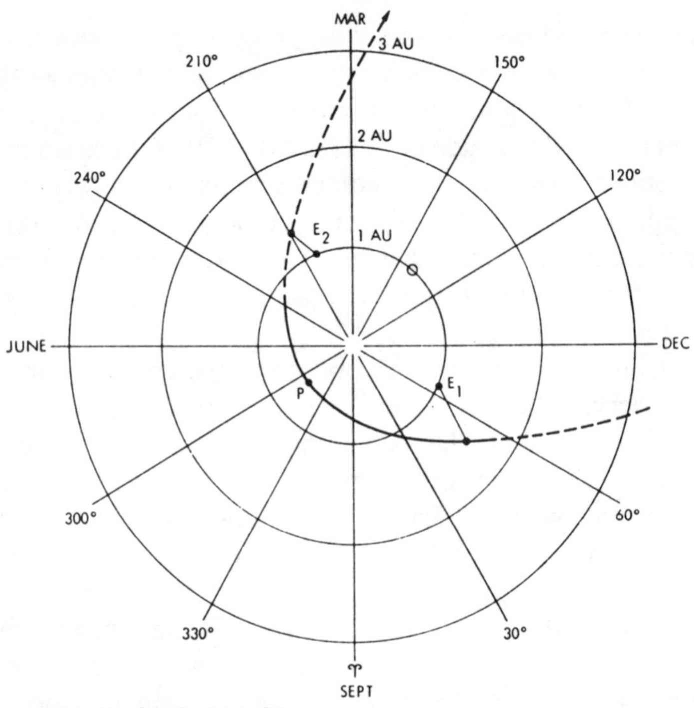

Fig. 10-2. Relative Positions of Halley's Comet and the Earth. During perihelion passage, the comet will be directly opposite the earth on the other side of the sun. Viewing conditions will be best near the times of closest approach of the comet to the earth.

the larger bodies of our solar system. This orbit is neither stable nor permanent. Halley's comet was initially "captured" when its velocity was decreased by the gravitational field of one of the larger planets, most probably Neptune, since it is a member of Neptune's "family" of comets. Halley's comet was thus deflected several thousand years ago from a long-period orbit into a short-period one, and future perturbations of its present elongated ellipse will occur as it again closely encounters one of the planets, especially Jupiter or Saturn.

Should its velocity be further slowed, a smaller elliptical orbit would result, which might possibly cause the comet to approach closer to the sun at perihelion and perhaps to orbit directly onto a collision course with the sun. On the other hand if, because of the relative motions of Halley's comet and the planet, the comet's velocity should be accelerated, then the orbital change may result in a hyperbola. Such a hyperbolic orbit would whip the comet out of our solar system and into the darkness of deep space, its blazing countenance extinguished forever.

In the year 1682, while observing the return of Halley's comet, Johannes Hevelius noted a luminous ray ejected from the nucleus. He drew a picture of this event which appeared in his *Annus Climactericus*, published in 1685. During the apparition of 1835, Friedrich Bessel, the German astronomer who predicted the existence of the planet Neptune, noted small "jets" protruding from the nucleus, and at the same time Captain William Henry Smyth (1788-1865), of the Royal Navy and Foreign Secretary of the Royal Astronomical Society, recorded a luminous area within the comet's nucleus. During the return of Halley's comet in 1910, violent processes occurring within its tail were noted and perhaps a separation of a small nuclear fragment from the main body was photographed (Fig. 10-3). All of this vigorous internal activity suggests a certain instability in the composition of the comet and portends a possible total disruption of the nuclear core of Halley's comet in the future.

When Comet West appeared in 1975, multiple splitting of the nucleus was observed. A total of four fragments that once comprised the former "solid" nucleus was noted. Only two other comets are known to have broken into more than four parts, while an additional dozen comets have been observed to separate into two or three components. Most comets, however, demonstrate a slower rate of disintegration.

Short-period comets tend to be dimmer than their long-period cousins, an apparent result of their frequent brief

Fig. 10-3. Halley's Comet on 6 June 1910. A detached portion of the comet is seen drifting away from the main body. Similar disruptive activity was witnessed during previous apparitions—signs of the unstable nature of the comet.

encounters with our star. During each perihelion approach, both dust and gas are lost from the nucleus by the action of the solar wind, light pressure, and gravitational effect of the sun. The coma and spectacular tail are in fact the sum of the comet's losses at any moment. The mass of the comet is being slowly reduced and matter is being dispersed as dusty debris into space. As the earth annually intersects the orbits of these disintegrating comets, some fragments rain upon our atmosphere, producing spectacular meteor showers.

The close association between comets and meteors was first established by the curious behavior of Biela's comet. Wilhelm von Biela (1782-1856) studied astronomy while still an officer in the Austrian army, and he eventually became a comet-hunter. In 1826, Biela rediscovered a comet previously observed in 1772. He calculated its orbit and found that it had a period of about six years. Because of this work his own name was assigned to the comet. Biela's comet is important because it was the first to confirm Kepler's suspicion of over two centuries before that comets "fall prey to Death." In 1845, the comet was observed to have divided in two, apparently having been disrupted by Jupiter when it passed the planet several years before. In 1852, the two parts of the comet were both very faint and separated by over a million miles. Before its next scheduled return, Biela himself was dead and his comet never again appeared. Instead, a swarm of meteors was observed with its radiant point in the constellation of Andromeda. These Andromedids or Bielids were the first strong evidence of the relationship between comets and meteors.

We now know that both the Eta Aquarid meteor shower in early May and the Orionid shower in October each year are related to Halley's comet. They are testimony of this comet's slow disintegration. Halley's comet may ultimately disperse into a swarm of meteors, some of which may in time encounter the earth and dazzle us briefly in tribute to what was once an object of both beauty and terror.

In 1937, a body about a half-mile across, later named Hermes, passed about 500,000 miles from the earth, less than twice the distance to the moon. It has not been seen again. The first such object known to cross the earth's orbit was discovered in 1932 and was named for the Greek solar deity Apollo. In the same year a small asteroid, Amor, was discovered with an orbit similar to Apollo's. Since then, many other objects with perihelions smaller than 1.3 astronomical units have been found. (One *astronomical unit* or *AU* is the distance from the earth to the sun or approximately 93 million miles.) They are all designated as Apollo-Amor objects. Since their orbits all range in the vicinity of the earth's, it is only a matter of time before such an object will strike the earth. According to an article by George W. Wetherill in *Scientific American* for March 1979, the statistical probability is that an Apollo object larger than a half-mile in diameter will collide with the earth once every 250,000 years. The impact would produce a crater many miles in diameter. Wetherill estimates that perhaps 1,500 Apollo-like objects have struck the earth during the past 600 million years.

The nuclei of the comets lie in the same size range as those of Apollo objects, and it is now hypothesized that these orbiting hunks of rock are not true asteroids but rather may be the outgassed remnants of comets. It was the Estonian astronomer Ernst Öpik who initially suggested that Apollo objects may be "extinct" comets. The scenario of this cometary transformation is a simple one. In the course of hundreds or thousands of perihelion passages, the nucleus of a short-period comet is steadily depleted of its volatile substances, while the denser rocky material remains behind. The activity of Comet Encke, for instance, has declined noticeably during its observed passages, and the comet may be well on its way to becoming inactive. Other periodic comets sighted in the past have shown little evidence of cometary activity of late and may also be approaching extinction. Such comets become "asteroidal" and their appearance is indistinguishable from an Apollo object.

Although the earth is frequently showered by the dust from disintegrating comets, the chance of a collision of this planet with a sizable chunk of comet matter is more remote. Yet there is abundant evidence imprinted upon the face of the earth of significant past collisions with celestial debris. In fact, a gigantic catastrophe occurred within this century when the earth impacted with a large meteorite, probably the remains of a comet.

On the morning of 30 June 1908, an immense explosion occurred above the Stony Tunguska River basin of remote Central Siberia. A bright fireball burst overhead, and twelve hundred square miles of relatively unpopulated forest was blown flat. Men and animals over thirty miles from the epicenter were seared by the heat. The sound of the explosion was heard for 600 miles. The shock wave it occasioned circled the globe twice. According to an article in an October 1981 issue of *Science*, up to 45% of the life-protecting ozone in the Northern Hemisphere may have been depleted, resulting in twice the amount of ultraviolet radiation striking the earth for the next three years.

Various causes for this "Tunguska Event" were proposed, ranging from a nuclear explosion to an encounter with antimatter or a miniature black hole. It is now generally believed that the earth collided with a piece of a comet. Unfortunately, its nuclear fragment disintegrated several miles above the ground, leaving few material clues behind for scientists to find. Slovak astronomer Lubor Kresak, writing in the *Bulletin of the Astronomical Institutes of Czechoslovakia* in 1978, presented strong evidence that this fragment was an outgassed remnant of Comet Encke, which had separated from the main body thousands of years previously. In the conclusion to his findings, Kresak writes, "The identification of the Tunguska object as an extinct cometary fragment appears to be the only plausible explanation of the event; and a common origin with comet Encke appears very probable."

In April of the year 837, during its post-perihelion passage, Halley's comet passed less than four million miles from the earth. Severe perturbations were suffered in the orbit of the comet because of this close encounter with the earth. Our planet, however, sustained no damage, though many of its inhabitants who witnessed the brilliant spectacle were undoubtedly terrified. But the next time our planet may not be quite so fortunate.

It is not inconceivable that in future years Halley's comet will suffer both a reduction in its orbital period and a decrease in its activity, gradually metamorphosing into an Apollo object. Virtually invisible, it will wander silently and eternally through the heavens—unless at some remote date it impacts with one of its own kind, the moon, or a planet. Should that planet be our own, the fear of numerous generations of anxious comet watchers that the comet might strike the earth would at last be realized, and Halley's belief that such a collision "might reduce this beautiful world to its original chaos" would be confirmed. A fitting end, perhaps, to both a comet and a planet which together have shared a remarkable and wonderful history.

APPENDIX A

Summary of Comet Halley

Observational History: With only a single exception, Halley's comet has been observed at every return since 240 B.C.

Classification: Short-period comet, designated P/Halley, 1982i.

Motion: Retrograde.

Mean Orbital Period: 76.8 years. The range is from 74 years and 5 months between the returns of 1835 and 1910 to 79 years and three months between the returns of 451 and 530. Nongravitational effects lengthen the orbital period by 4.1 days at each apparition.

Aphelion: 1948.

Perihelion: 9 February 1986 at a distance of 54.6 million miles from the sun. For this return, the comet was first sighted on 16 October 1982 when it was more than a billion miles from the sun.

Previous Perihelion: 20 April 1910.

Next Perihelion: 29 July 2061.

Velocity at Perihelion: Approximately 34 miles per second or 122,000 miles per hour.

Closest Approach to The Earth: 11 April 1986 at a distance of 39 million miles.

Closest Known Approach to The Earth: 10 April 837 at a distance of less than 4 million miles.

Orbit Description: An elongated ellipse with a major axis of 35.6 AU, a minor axis of 9 AU, and an eccentricity of 0.967.

Angular Elements of the Orbit:

Inclination 162 degrees

Argument of perihelion 112 degrees

Longitude of ascending node 58 degrees

Brightness: An exceptionally bright, short-period comet. Its intrinsic brightness will increase following perihelion. However, its unfavorable position referable to the sun and earth during this apparition will limit it to a magnitude 3 or 4 object during March and April 1986.

Nucleus: A conglomerate of solid meteoric material and frozen gases with a diameter of approximately three miles.

Tail: Two well-developed gas and dust tails which may reach 40 degrees in length following perihelion.

Associated Meteor Showers: The Eta Aquarids in early May and the Orionids in late October.

Optimal Viewing Times: In the southern sky during the evenings of December 1985 and January 1986 and the mornings of March and April 1986.

Some Noted Persons Associated With Past Apparitions: Flavius Josephus, Attila the Hun, William the Conqueror, Giotto di Bondone, Pope Calixtus III, Johannes Kepler, Isaac Newton, Alexis Clairaut, Charles Messier, Mark Twain.

APPENDIX B

Track of Comet Halley 1974 - 1986

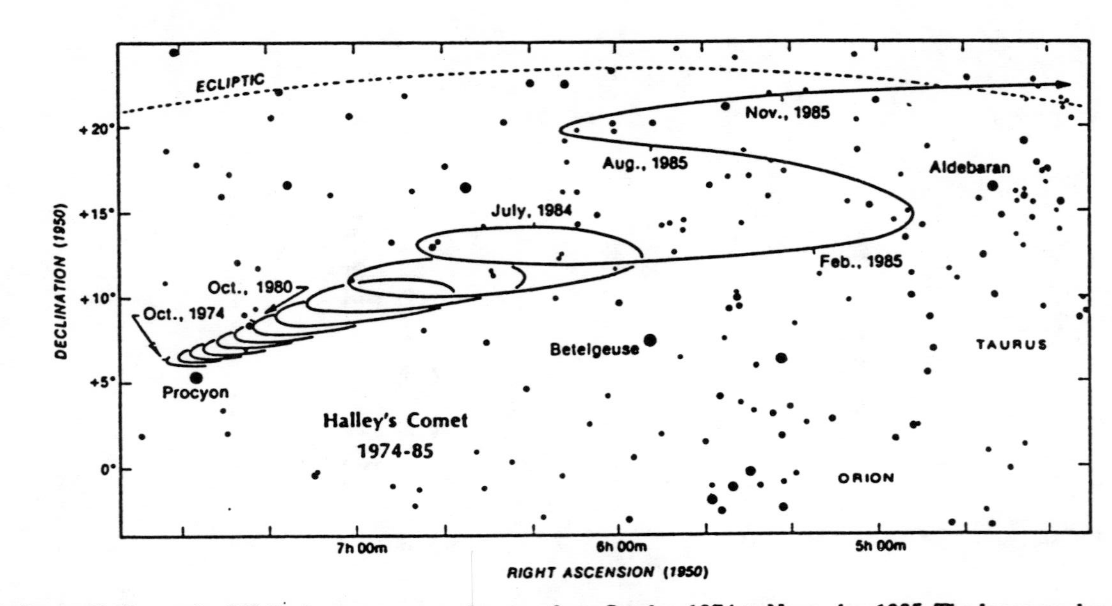

Fig. B-1. Pre-perihelion path of Halley's comet among the stars from October 1974 to November 1985. The loops are due to the orbital motion of the earth.

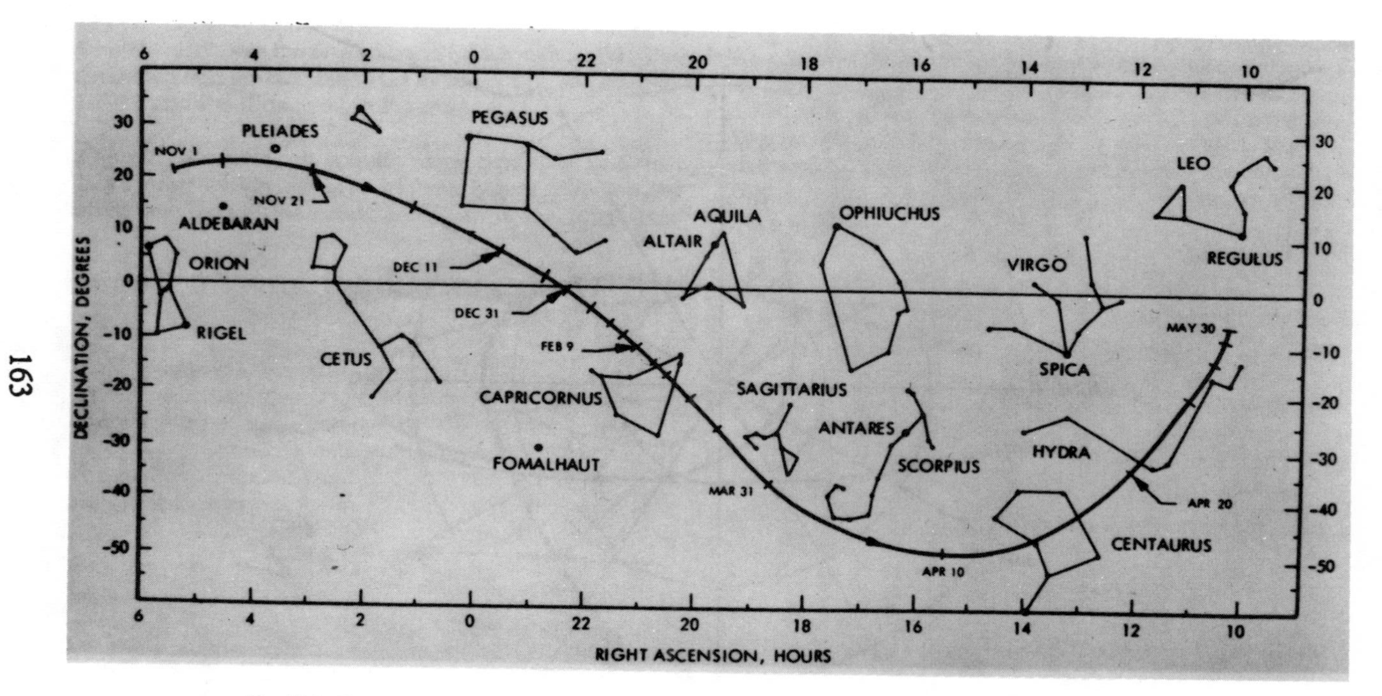

Fig. B-2. Path of Comet Halley through the constellations from November 1985 to May 1986.

APPENDIX C

Sky Positions And Magnitudes of Comet Halley In 1986

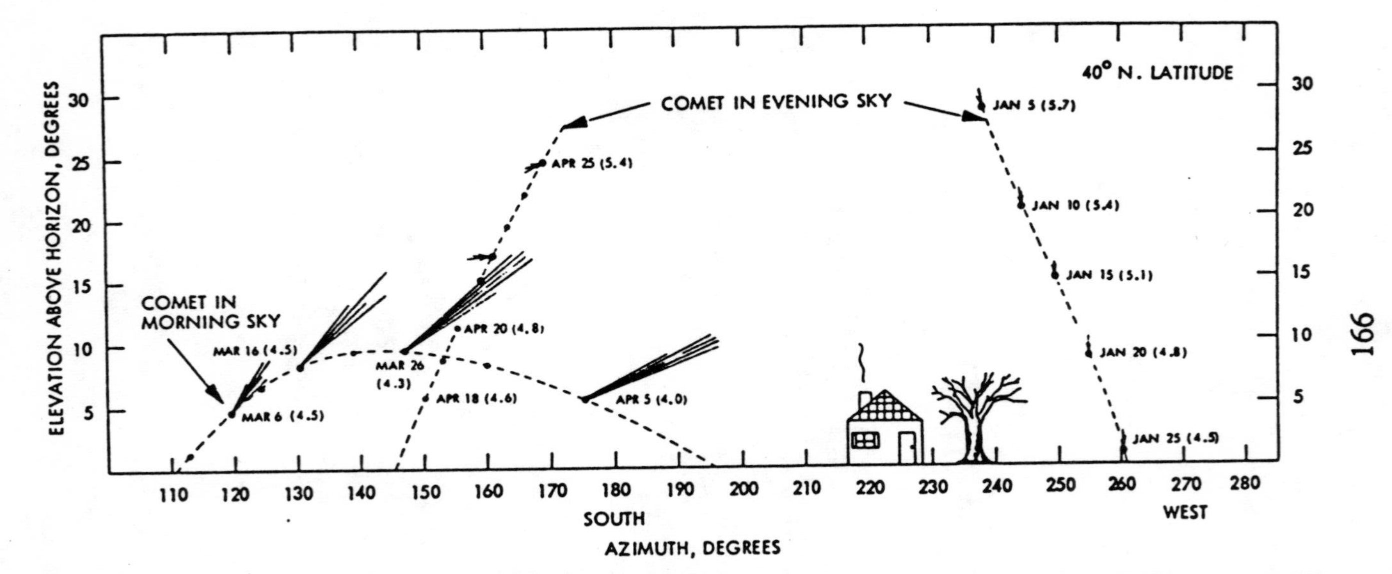

Fig. C-1. Observers Located at 40 degrees North Latitude. The cities of Denver, Philadelphia, Madrid, and Rome lie near this geographical latitude. Comet positions are for times about 90 minutes after sunset or about 90 minutes before sunrise. The apparent total magnitude for each date is shown in parentheses. The tail length is approximate.

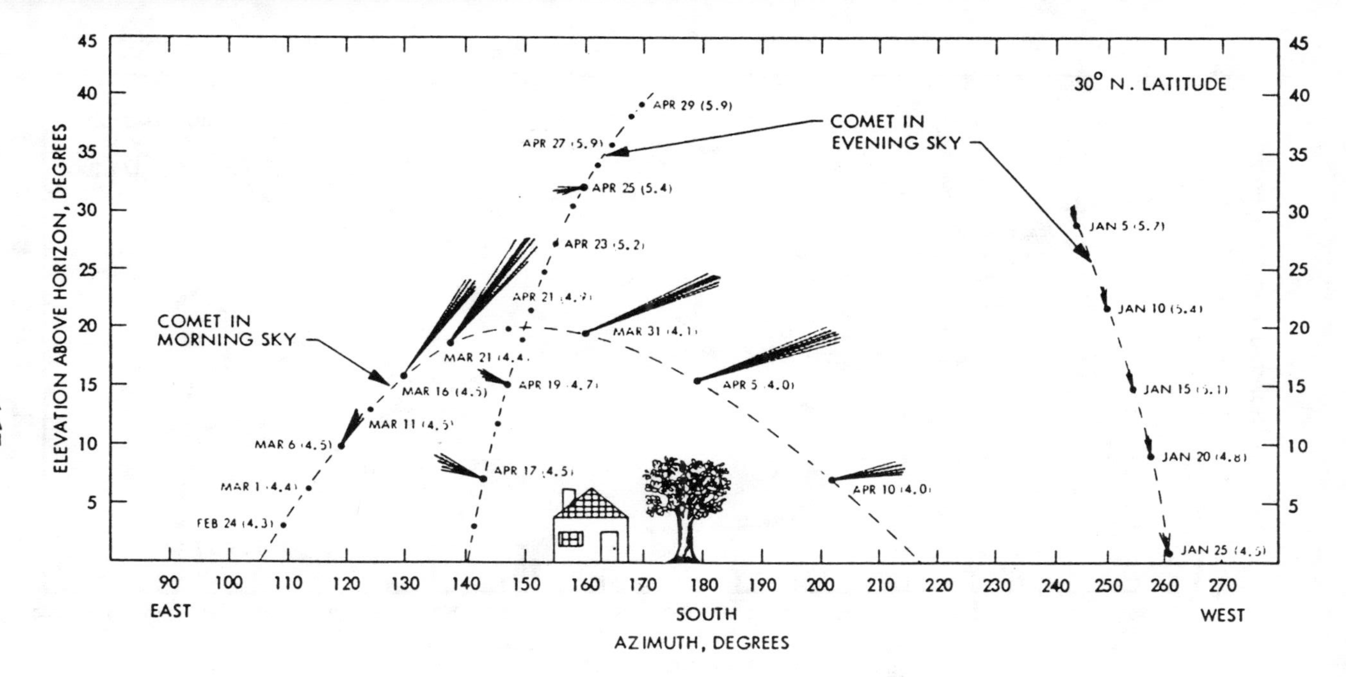

Fig. C-2. Observers Located at 30 degrees North Latitude. The cities of Houston, New Orleans, Cairo, and Shanghai lie near this geographical latitude. Comet positions are for times about 90 minutes after sunset or about 90 minutes before sunrise. The apparent total magnitude for each date is shown in parentheses. The tail length is approximate.

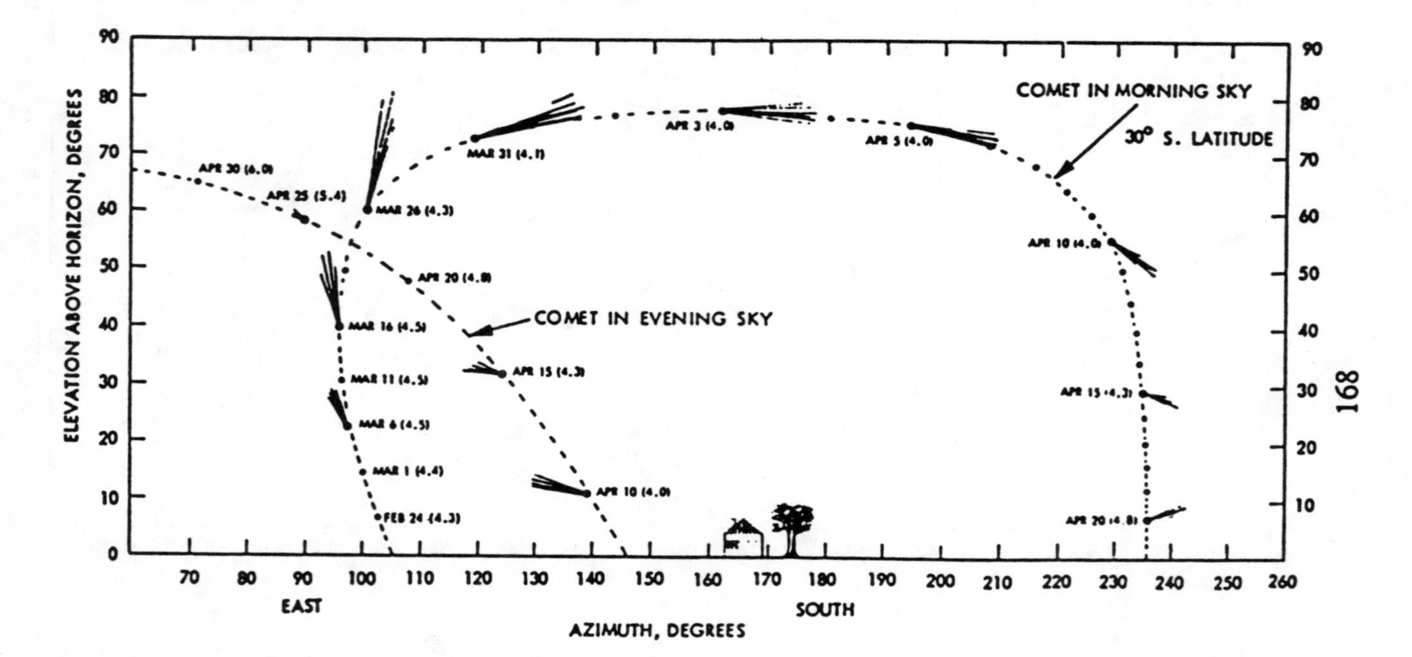

Fig. C-3. Observers Located at 30 degrees South Latitude. The cities of Perth and Sydney, Australia, and Cape Town, South Africa, lie near this geographical latitude. Comet positions are for times about 90 minutes after sunset or about 90 minutes before sunrise. The apparent total magnitude for each date is shown in parentheses. The tail length is approximate.

GLOSSARY

Aphelion: the point of a comet's orbit that is farthest from the sun.

Apollo body: a type of asteroid, quite possibly the remnant of an extinct comet nucleus, that has its perihelion inside the orbit of the earth.

Apparition: the appearance of a periodic comet.

AU: the astronomical unit, which is equal to the average distance between the earth and the sun, or approximately 93 million miles.

Capture: when a comet passes close to one of the larger planets, it may be so influenced by the planet's gravitational attraction that its orbit is changed into an ellipse and it becomes a member of that planet's "family" of comets.

Coma: the nebulous envelope that surrounds a comet's nucleus and forms the visible part of its head.

Conic section: one of three curves—ellipse, parabola, or hyperbola—which the orbit of a comet can assume.

Conjunction: the close approximation of two or more celestial bodies as viewed from the earth. A conjunction theory has been proposed as an explanation of the star of Bethlehem.

Declination: analogous to geographical latitude; measured in degrees north or south of the celestial equator.

Ecliptic: the plane of the earth's orbit about the sun on which eclipses occur. A comet's inclined orbit intersects the ecliptic at two points called the nodes.

Elements of an orbit: the six quantities necessary to define the exact nature of a comet's orbit. These are the argument of perihelion, longitude of the ascending node, inclination, perihelion distance, time of perihelion passage, and eccentricity.

Ephemeris: a listing of the positions of a comet in the firmament over regular intervals of time.

Magnitude: the apparent brightness of a comet or other celestial body as viewed by the unaided eye. An object of magnitude 6 is just visible. The magnitude of Venus is -4 and Jupiter is -2, while Halley's comet should be 3 or 4 at its brightest. Comet Halley is intrinsically brighter during the post-perihelion part of its orbit.

Meteor: the light phenomenon observed during the passage of a solid body through the earth's atmosphere from outer space.

Meteorite: a meteor that has landed on the earth and still retains its identity as being interplanetary in origin.

Node: one of two opposite points where the orbit of a comet intersects the ecliptic, designated as ascending node or descending node.

Nongravitational effect: the emission of volatile gases from the nucleus of a comet acts to cause the comet to deviate from its predicted perihelion time.

Nucleus: the small, brighter, and denser portion of the head of a comet, which is surrounded by the coma.

Oort's cloud: the postulated cloud of comets which surrounds the solar system and is under the gravitational influence of the sun.

Perihelion: the point of a comet's orbit that is nearest to the sun.

Period: the time required for a comet to complete one revolution of its orbit. A short-period comet like Halley's has a period of less than two-hundred years; each revolution of a long-period comet exceeds two centuries.

Perturbation: the disturbance in the elliptical orbit of a comet by the gravitational attraction of the planets.

Retrograde motion: motion in a comet's orbit opposite to the usual orbital direction of the planets.

Right ascension: analogous to geographical longitude; measured eastward from the vernal equinox in hours, minutes, and seconds.

Solar wind: the stream of charged particles ejected from the sun and responsible for the ion tails of comets.

Syzygy: an alignment of three celestial bodies. During the apparition of 1910, Halley's comet was aligned with the earth and the sun.

Tail: extending for millions of miles in an antisolar direction, this is a comet's most identifiable feature. Comets may possess both ion tails and dust tails.

Zodiacal light: a glow in the night sky caused by sunlight reflecting off interplanetary dust particles which have been deposited by comets.

Selected Bibliography

Chapter One

1. Brown, Peter L. "Comets in History." Chap. 1 in *Comets, Meteorites and Men*. New York: Taplinger, 1974.
2. Chambers, George F. "Comets in History and Poetry." Chap. 14 in *The Story of the Comets*. Oxford: Clarendon Press, 1909.
3. Gingerich, Owen. "Tycho Brahe and the Great Comet of 1577." *Sky and Telescope* 54 (December 1977): 452-58.
4. Ley, Willy. "Dr. Edmond Halley and the 'Year of the Comet.' " Chap. 7 in *Watchers of the Skies* . New York: Viking Press, 1963.
5. Whitney, Charles A. "The Sky Awakes." Chap. 3 in *The Discovery of Our Galaxy*. New York: Alfred A. Knopf, 1971.

Chapter Two

1. Armitage, Angus. *Edmond Halley*. London: Thomas Nelson and Sons, 1966.
2. Bobrovnikoff, Nicholas T. "Edmond Halley, 1656-1742." *Scientific Monthly* 55 (November 1942): 438-46.
3. MacPike, Eugene F., ed. *Correspondence and Papers of Edmond Halley*. Oxford: Clarendon Press, 1932.
4. Ronan, Colin A. *Edmond Halley, Genius in Eclipse*. Garden City: Doubleday and Company, 1969.

Chapter Three

1. Eddington, A.S. "Halley's Observations on Halley's Comet, 1682." *Nature* (London) 83 (26 May 1910): 372-73.
2. Ley, Willy. *Visitors From Afar: The Comets*. New York: McGraw-Hill, 1969.
3. Plummer, H.C. "Halley's Comet and Its Importance." *Nature* (London) 150 (29 August 1942): 249-57.

Chapter Four

1. Doolittle, C.L. "Halley's Comet." *Popular Science Monthly* 76 (January 1910): 5-22.
2. Heward, Edward Vincent. "The Story of Halley's Comet." *Nineteenth Century and After* 66 (September 1909): 509-26.
3. Itard, Jean. "Alexis-Claude Clairaut." In *Dictionary of Scientific Biography*, edited by Charles Coulston Gillespie, vol. 3. New York: Charles Scribner's Sons, 1971.
4. "The Approaching Comet." *Edinburgh Review* 61 (April 1835): 82-128.
5. Watson, James C. *Treatise on Comets*. Philadelphia: James Challen and Son, 1861.

Chapter Five

1. Airy, George B. "An Address on Presenting the Honorary Medal to Professor Rosenberger." *Memoirs of the Royal Astronomical Society* 10 (1838): 376-89.
2. Brady, Joseph L. "The Effect of a Trans-Plutonian Planet on Halley's Comet." *Publications of the Astronomical Society of the Pacific* 84 (April 1972): 314-22.
3. Chambers, George F. "Halley's Comet." Chap. 9 in *The Story of the Comets*. Oxford: Clarendon Press, 1909.
4. Hind, J. Russell. "The Comet of Halley." Chap. 4 in *The Comets*. London: John W. Parker and Son, 1852.
5. Kiang, T. "The Cause of the Residuals in the Motion of Halley's Comet." *Monthly Notices of the Royal Astronomical Society* 162 (1973): 271-87.
6. Proctor, Mary and A.C.D. Crommelin. "The Story of Halley's Comet." Chap. 4 in *Comets*. London: Technical Press, 1937.

Chapter Six

1. Cowell, P.H., and A.C.D. Crommelin. "The Perturbations of Halley's Comet in the Past." First through Fifth Papers. *Monthly Notices of the Royal Astronomical Society* 68:111-25, 173-79, 375-78, 510-14, 665-70.
2. Hind, J.R. "On the Past History of the Comet of Halley." *Monthly Notices of the Royal Astronomical Society* 10(11 January 1850): 51-58.
3. Ho Peng Yoke. "Ancient and Mediaeval Observations of Comets and Novae in Chinese Sources." *Vistas in Astronomy* 5 (1962): 127-225.

4. Kiang, T. "The Past Orbit of Halley's Comet." *Memoirs of the Royal Astronomical Society* 76,pt.2 (1972): 27-66.
5. Yeomans, Donald K., and Tao Kiang. "The Long-term Motion of Comet Halley." *Monthly Notices of the Royal Astronomical Society* 197 (1981): 633-46.

Chapter Seven

1. Cortie, A.L. "The Devil, the Turk, and the Comet." *Observatory* 33 (February 1910): 91-95.
2. Flammarion, Camille. "Comets in Human History." Chap. 1 of Book 5 in *The Flammarion Book of Astronomy*. New York: Simon and Schuster, 1964.
3. Freitag, Ruth S., comp. *The Star of Bethlehem: A Selected List of References*. Library of Congress, 1978.
4. Olson, Roberta J.M. "Giotto's Portrait of Halley's Comet." *Scientific American* 240 (May 1979): 160-70.
5. Rigge, William F. "An Historical Examination of the Connection of Calixtus III with Halley's Comet." *Popular Astronomy* 18 (April 1910): 214-19.
6. Warner, Irene E. Toye. "Great Events in the World During Apparitions of Halley's Comet." *Knowledge & Scientific News* 6 (December 1909): 463-66.

Chapter Eight

1. Alter, Dinsmore. "Comets and People." *Griffith Observer* 20 (July 1956): 74-82.
2. "Could the Earth Collide with a Comet?" *Scientific American* 102 (5 March 1910): 194.
3. Kaempffert, Waldemar. "The Most Famous of Comets." *Collier's* 45 (2 April 1910): 21-22.
4. Klein, Jerry. "When Halley's Comet Bemused the World." *New York Times Magazine* 8 May 1960.
5. McAdam, D.J. "The Menace in the Skies. The Case for the Comet." *Harper's Weekly* 54 (14 May 1910): 11-12.
6. Oppenheimer, Michael, and Leonie Haimson. "The Comet Syndrome." *Natural History* 89 (December 1980): 54-61.
7. Stephenson, Bill. "The Panic Over Halley's Comet." *MacLean's* 14 May 1955.
8. "The Passing of Halley's Comet." *Christian Herald* 1 June 1910, 519.

Chapter Nine

1. Biermann, Ludwig F., and Rhea Lüst. "The Tails of Comets." *Scientific American* 199 (October 1958): 44-50.
2. Oort, J.H. "The Structure of the Cloud of Comets Surrounding the Solar System, and a Hypothesis Concerning Its Origin." *Bulletin of the Astronomical Institutes of the Netherlands* 11 (13 January 1950): 91-110.
3. Washburn, Mark. "In Pursuit of Halley's Comet." *Sky and Telescope* 61 (February 1981): 111-13.

4. Whipple, Fred L. "A Comet Model." Parts 1-3. *Astrophysical Journal* 111:375-94; 113:464-74; 121:750-70.
5. Whipple, Fred L. "The Nature of Comets." *Scientific American* 230 (February 1974): 48-57.
6. Whipple, Fred L. "Background of Modern Comet Theory." *Nature* 263 (2 September 1976): 15-19.

Chapter Ten

1. Roosen, Robert G. and Brian G. Marsden. "Observing Prospects for Halley's Comet." *Sky and Telescope* 49 (June 1975): 363-64.
2. Sekanina, Zdenek. "Disintegration Phenomena in Comet West." *Sky and Telescope* 51 (June 1976): 386-93.
3. "The Tunguska Event and Encke's Comet." *Sky and Telescope* 56 (December 1978): 497-98.
4. Turco, R.P., O.B. Toon, C. Park, R.C. Whitten, J.B. Pollack, and P. Noerdlinger. "Tunguska Meteor Fall of 1908: Effects on Stratospheric Ozone." *Science* 214 (2 October 1981): 19-23.
5. Wetherill, George W. "Apollo Objects." *Scientific American* 240 (March 1979): 38-49.
6. Yeomans, Donald K. *The Comet Halley Handbook*. 2d ed. Pasadena: NASA, Jet Propulsion Laboratory, 1983.

INDEX

Page numbers in *italics* refer to captions for illustrations

ILLUSTRATION CREDITS

Page 5, courtesy of the American Museum of Natural History; 8, 10, Yerkes Observatory; 12, 13, from *Comets* by David C. Knight. Copyright © 1968 by Franklin Watts, Inc. Used by permission of the publisher; 18, The Royal Society; 20, Yerkes Observatory; 23, The Royal Society; 25, 28, Yerkes Observatory; 34, Chambers' *The Story of the Comets*, 1909; 36, courtesy of J. Classen and Pulsnitz Observatory; 37, *Memoirs of the Royal Astronomical Society*, 9 (1836); 38, Royal Greenwich Observatory; 41, diagram by author; 43, NASA and Donald K. Yeomans; 53, 55, Archives de l'Académie des Sciences de Paris; 61, 76, 78, Yerkes Observatory; 79, Mount Wilson and Las Campanas Observatories, Carnegie Institution of Washington; 89, National Portrait Gallery, London; 96, Royal Astronomical Society; 98, The Royal Society; 99, Royal Astronomical Society; 111, Alinari/Art Resource, NY; 114, with special authorization of the City of Bayeux; 116, master and Fellows of Trinity College, Cambridge; 117, *Popular Science Monthly*, 76 (1910); 120, Rare Books and Manuscripts Division, The New York Public Library, Astor, Lenox and Tilden Foundations; 125, courtesy of the American Museum of Natural History; 127, 128, Yerkes Observatory; 134, courtesy of *Christian Herald;* 135, courtesy of the American Museum of Natural History; 139, diagrams by author; 150, Palomar Observatory/California Institute of Technology; 152, NASA and Donald K. Yeomans; 154, Yerkes Observatory; 162, *Sky & Telescope* diagram by Roger Sinnott, used by permission. © 1975 Sky Publishing Corp.; 163, 166, 167, 168, NASA and Donald K. Yeomans.